THE GAS TRAMCAR

THE GAS TRAMCAR

AN IDEA AHEAD OF ITS TIME

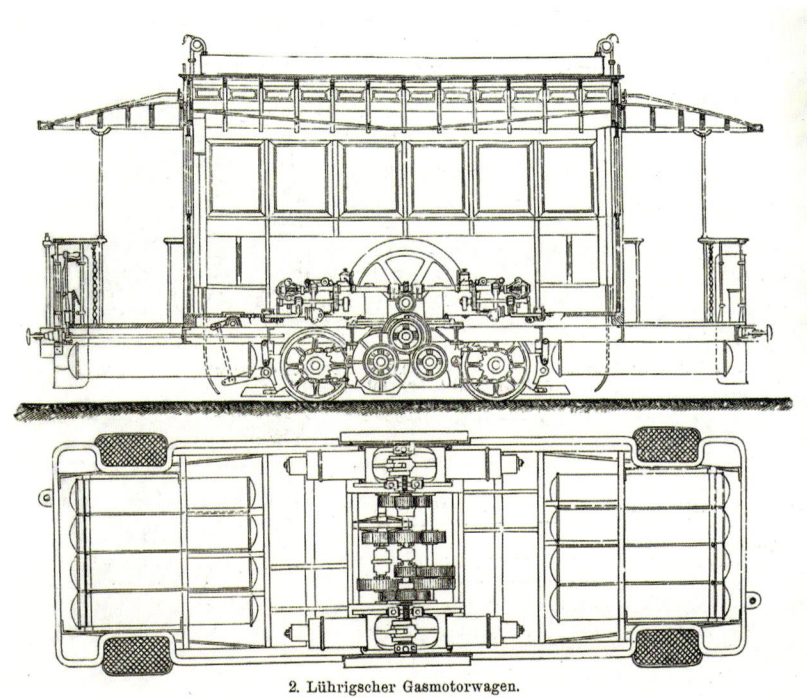

2. Lührigscher Gasmotorwagen.

JOHN HANNAVY

AN IMPRINT OF PEN & SWORD BOOKS LTD.
YORKSHIRE – PHILADELPHIA

First published in Great Britain in 2022 by
Pen & Sword Transport
An imprint of
Pen & Sword Books Ltd
Yorkshire - Philadelphia

Copyright © John Hannavy, 2022

ISBN 978 1 39909 601 0

The right of John Hannavy to be identified as Author of this work has been asserted by him in accordance with the Copyright, Designs and Patents Act 1988.

A CIP catalogue record for this book is available from the British Library.

All rights reserved. No part of this book may be reproduced or transmitted in any form or by any means, electronic or mechanical including photocopying, recording or by any information storage and retrieval system, without permission from the Publisher in writing.

Typeset in Palatino 10/13
by SJmagic DESIGN SERVICES, India.
Printed and bound by Printworks Global Ltd,
London/Hong Kong.

Pen & Sword Books Ltd incorporates the Imprints of Pen & Sword Books Archaeology, Atlas, Aviation, Battleground, Discovery, Family History, History, Maritime, Military, Naval, Politics, Railways, Select, Transport, True Crime, Fiction, Frontline Books, Leo Cooper, Praetorian Press, Seaforth Publishing, Wharncliffe and White Owl.

For a complete list of Pen & Sword titles please contact

PEN & SWORD BOOKS LIMITED
47 Church Street, Barnsley, South Yorkshire, S70 2AS, England
E-mail: enquiries@pen-and-sword.co.uk
Website: www.pen-and-sword.co.uk

or

PEN & SWORD BOOKS
1950 Lawrence Rd, Havertown, PA 19083, USA
E-mail: Uspen-and-sword@casematepublishers.com
Website: www.penandswordbooks.com

Author's website: www.johnhannavy.co.uk

By the same author, also published by Pen & Sword:
TRANSPORTER BRIDGES – *an illustrated history*
THE GOVERNOR – *controlling the power of Steam Machines*

Cover image: Sectional diagrams of the engine and gearing on one of the smaller Blackpool trams. Insets: Neath Corporation Tramways Car No.1 at Cefn Coed Museum, and an Edwardian postcard of Car No.6 at Skewen.

Title page image: One of Carl Lührig's original designs for his proposed larger gas tramcars, with the gas tanks installed longitudinally. From an illustration in the 14th Edition of Brockhaus' Konversations-Lexikon, published in 1896.

Contents page image: A sectional view of the engine in Lührig's 'Gas-Motor Car', as published in Frank H. Mason's 'Gas-Motor Street Cars' in the June 1895 edition of the American publication *Cassier's Magazine*. Here the gas tanks are installed transversely. Mason, a career diplomat, had been appointed the US Consul-General in Frankfurt in 1889 and regularly published articles on European technology.

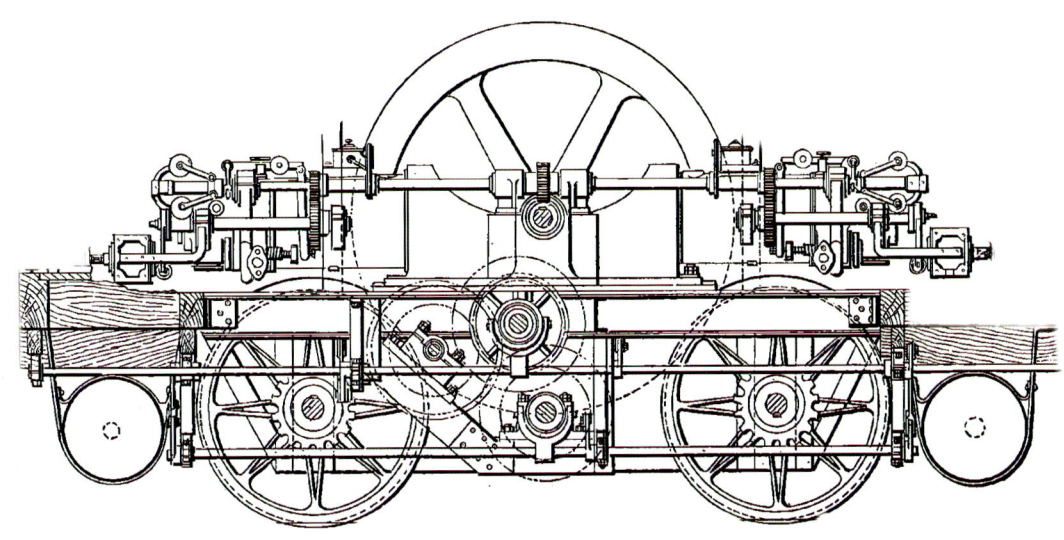

CONTENTS

INTRODUCTION	7
THE ROUGH ROAD TO PROGRESS	11
RUNNING ON RAILS	17
RAISING STEAM	27
BATTERIES, CONDUITS, CABLES, OIL AND COMPRESSED AIR	37
THE GAS REVOLUTION	53
THE FIRST GAS TRAMCARS	69
THE GAS TRAMCAR IN BRITAIN	111
ELECTRICITY WINS THE DAY	165
CAN THERE BE A FUTURE FOR GAS-ENGINED TRAMCARS?	173
SELECTED PATENTS 1791–1903	180
A CHOICE OF GASES	196
BIBLIOGRAPHY	198
ACKNOWLEDGEMENTS	200
INDEX	201

INTRODUCTION

A rare vehicle even when new – and now the only one of its kind anywhere in the world – is not something you would normally expect to find tucked away in the corner of a colliery museum, and yet that was the case when we travelled to south-west Wales to the Cefn Coed Colliery Museum near Neath.

Sometimes the original reason for visiting a place turns out not to be what forms the enduring memories of the trip. My visit to Cefn Coed had been planned around seeing the huge winding engine which had raised and lowered the cages at what was once the deepest anthracite mine in the country. The horizontal duplex winding engine was built in 1927 in Wigan by the Worsley Mesnes Company, which could trace its origins back to 1850 when it had been established as Worsley Mesnes Ironworks. The engine, which is currently turned by electricity for the benefit of visitors, is fitted with an uncommon centre-weighted Hartnell-type centrifugal governor with cylindrical rather than spherical weights, and that is what had put it on my 'must see' list while carrying out research for my book *The Governor – controlling the power of steam machines* which was published by Pen & Sword in 2021.

Ours was the only vehicle in the car park on that cold August morning as we made our way to the colliery buildings in driving rain. We had intentionally arrived at opening time, which usually means I get the place to myself before the crowds turn up. By the time I had photographed the engine and its governor from every angle, there was still nobody else in sight – hardly surprising, given the weather.

Having completed the fascinating self-guided tour of the colliery – and enjoyed a nice cup of tea and a chat with the lady in the café – that just left the museum's unique treasure to be investigated: the world's only surviving example of a gas-powered tram, albeit now without its engine. The tramcar had been rediscovered in the 1980s being used as a garage, before being rescued and restored to its present condition by local apprentices as part of the government's Youth Training Scheme.

The gas tram was an obvious, if ultimately short-lived, outcome of the emerging technologies of Victorian times and as with so many others, it was not the work of a single inventor. Indeed, a succession of engineers in Europe, Australia, New Zealand and the United States all believed they had created the first such vehicle – applying for, and being granted, numerous patents for so doing.

Opposite: The front of the only surviving Neath Corporation Tramways gas tram, displayed at Cefn Coed Colliery Museum. The tramcar was bought secondhand from Blackpool, one of the original batch of six small cars built by The Ashbury Railway Carriage and Iron Company. The engine space is now empty apart from the large flywheel. Most of Neath's fleet consisted of the larger cars built at the Lancaster Carriage Works.

Below left: The recreation of an underground tunnel at the Cefn Coed Colliery Museum.

Below right: The giant steam winding engine, built in Wigan, which once provided the colliery's power.

8 • THE GAS TRAMCAR

A rare photograph of Dessau Car No.6, Carl Lührig's design for a small gas-motor tram, seen here with the motor inspection and service doors open. A woodcut based on this photograph is illustrated on page 73. These cars entered service in Dessau on the route between the Post Office and the city's largest cemetery, Friedhof III.

In his company's 1896 promotional booklet *De Gasmotortram volgens het Systeem Lührig*, an un-named author working on behalf of 'The Gas Traction Company departm, Netherlands and its Colonies' based in Amsterdam wrote:

'But the idea soon developed of placing the [gas] engine on the vehicle itself, instead of on a locomotive, and a great number of inventors in all civilized lands sought to realise this idea; in Germany this was tested by Blessing, Capitaine, Daimler, the Körting brothers, Stevens and others; in England by Holt, Dyson, Nichols & March, Piers, McNay & Harrison, etc.; in France by Montclar; in Austria it was Lobenhofr and Anibas, in Italy Morani, in Australia Barnes & Danks, in America Connelly, etc., who suggested it.'

There were many more who foresaw the future of the gas engine, and while some of the names listed above have faded into obscurity, others went a long way towards seeing their dreams of gas traction realised.

Amazingly the number of patent applications for gas trams, their engines and their friction clutches between 1878 and 1920 surpassed the number of gas-engined tramcars actually operating with fare-paying passengers between those dates.

For a few years, gas trams were expected to offer a great, clean future – and a less costly one – but they were rapidly overtaken by other, more expensive but ultimately more practical, systems. There were also some proposed solutions which turned out to be much less practical.

An illustration of the same car, drawn for the American *Cassier's Magazine* and featured in its June 1895 issue, this time with the engine doors closed.

Gas tramcars were never common. Less than a handful of British tramway companies ever tried the vehicles out, only three brought them into regular service and fewer than three dozen such vehicles ever ran on British lines – and those accounted for almost half of the world's total.

Their limitation was that, in the early days of the internal combustion engine, those available were seen as not being powerful enough for the task of hauling tramcars laden with passengers over anything but the most level of terrains. They were also unfairly labelled as unreliable, and yet, when the last gas tramcars were withdrawn in 1920, they were over twenty years old.

Their window of opportunity came too early and passed before the full potential of the gas engine could be realised and developed.

This book explores how they came about, looks at their history within both the chronology of tramways and the development of the internal combustion engine, and anticipates a possible future for gas power.

It has been surprising to find how much of what has subsequently been written about these unusual vehicles – sometimes seen through the distorting mirror of hindsight – is at odds with contemporary reports and the surviving primary source material which has been unearthed in the course of this project.

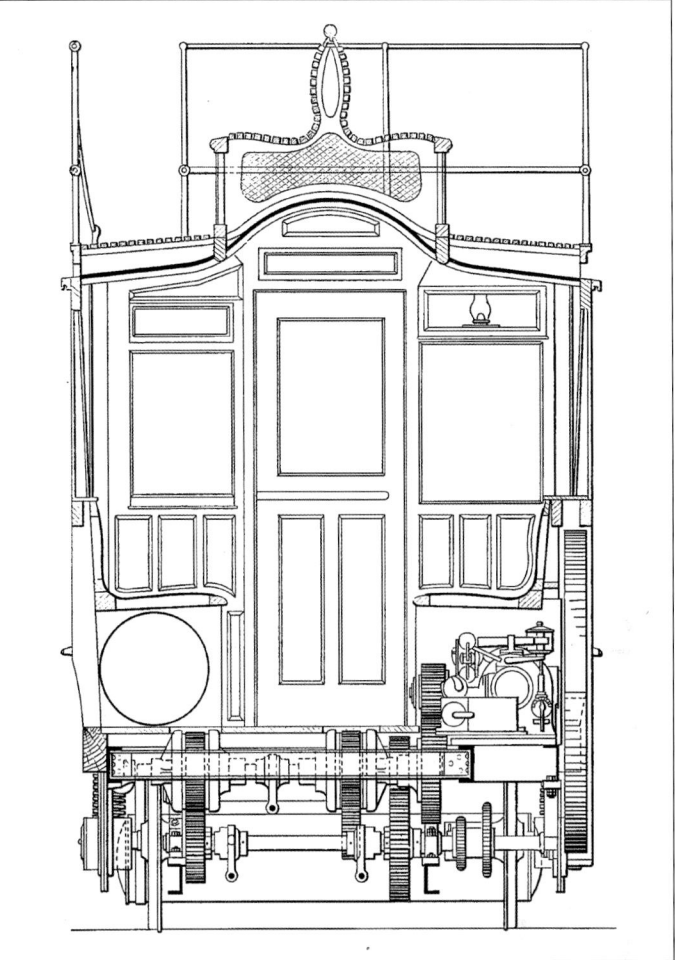

As more and more original material is catalogued and digitised in the world's archives, who knows how many other inventors might become associated with the novel idea of the gas tram?

Was the gas-engined tram simply an idea ahead of its time – more than a century ahead of its time in fact – and will gas-fuelled vehicles become the norm in the future as cleaner and environmentally greener transport systems are brought on stream?

Tramcars are enjoying renewed popularity worldwide as the need to improve air quality in city centres – currently heavily polluted by diesel and petrol engines – rises to the top of urban agendas. The present system of overhead catenary and electric vehicles is certainly cleaner than cars or buses, but when the carbon impact of both manufacture and power generation are taken into account, a twenty-first century gas tram may turn out to be both greener and cheaper to operate.

Gas railcars and tramcars using LNG or 'green' compressed biomethane are currently in development in the UK and several other countries, while others using clean hydrogen fuel cells to generate electricity rather than actually burning the gas are already running in China, Korea, Dubai, the Dutch Caribbean island of Aruba and elsewhere, so, it may be that a new age of clean and efficient gas trams could dawn in the near future, rather than them being no more than a memory from a century ago. The last chapter of this book explores that possibility.

John Hannavy 2022

The gearing of the 1894 double-decked tram, similar to those ordered from the Ashbury Railway Carriage and Iron Works for the proposed service on the Croydon and Thornton Heath tramway. The three gas tanks were located beneath the seats on the left side of the vehicle and beneath the end platforms. The motor and flywheel were to the right. Note that the bench seating on the open upper deck runs the length of the vehicle rather than across as was the case in the production version. Beneath that seat was a water tank, used to cool the gas engine. For a more detailed view of the engine and gearing from the opposite end of the vehicle, see page 109.

THE ROUGH ROAD TO PROGRESS

Travelling by public transport 200 years ago must have been an extremely uncomfortable experience. The rough and pitted road surfaces and the rocking and juddering of the iron-tyred horse-drawn coaches and omnibuses must have made for very much less than enjoyable journeys.

Written in her diary – later published as a book in 1874 – Dorothy Wordsworth's account of her travels through Scotland with her brother William and Samuel Taylor Coleridge in 1803 contains some unexpected commentaries on travel in the early nineteenth century, including an appendix in which she classified the Scottish roads along which they journeyed.

Her classifications ranged from 'excellent' for the road from Dalmally to Taynuilt, to 'wretchedly bad' for the route from Blair to Fascally. Between those extremes are such descriptions as 'tolerable', 'roughish', 'baddish', 'middling' and 'good'. Sadly for us, her criteria for each classification remained entirely within her own head, but those rated 'good' were few and far between, found only close to major towns and cities.

In 1819, the engineer Thomas Telford was working for the Roads Commissioners who sought to improve the infrastructure across the country to cater for the growing numbers of vehicles using the pitted and pot-holed roads. While travelling in Scotland with the Poet Laureate Robert Southey, however, Telford was surprised to discover that neither he nor his roads were welcome in one county along the way.

Writing in his *Journal of a Tour in Scotland,* Southey reported that he and Telford met some travellers coming south, 'from the fine new roads in the North of Scotland which are the best in the world' – which Telford had been responsible for, of course – who were shocked when their horses stumbled and the coach jolted so violently they were almost thrown out. 'What's the matter?' asked one of them. 'Perthshire – we're in Perthshire, Sir' replied their driver, as if that statement needed no further explanation.

Opposite: The Beach Hotel in Seaton, Devon, ran its own private omnibus which met holidaymakers from the railway station and transported them to the hotel. Travelling in such a vehicle must have been a terrifying start to a holiday. This postcard dates from c.1908.

Below: Horse omnibuses at Muirtown Locks near Inverness c.1910 await passengers about to arrive on a MacBrayne steamer having travelled through the Caledonian Canal.

'The cause of this mishap', wrote Southey when their own coach suffered the same fate, 'lay in the obstinacy of the Perthshire people, Perthshire being the only Highland county where they will not let the commissioners interfere with the management of their roads.'

There were numerous blacksmiths positioned at regular distances all the way along the route who made a reasonable living out of repairing damaged carriage wheels. It was they who apparently disapproved of the new roads that were being laid as they were losing a great deal of trade, with several even being put out of business as a result.

Where there were good roads, they had mostly been laid by Telford, who had been creating a smoother surface using crushed stone since 1801. His methods were later refined by the Scots engineer John Loudon Macadam who realised that smaller stones tightly impacted and cambered from the centre created road surfaces which were both more durable and better at dispersing water. These were not good roads by today's standards, but a huge improvement on what had gone before.

Macadam also discovered that the size of the stones on the surface was crucial if the 4-inch wide wheels on early nineteenth century wagons were not going to easily break it up, and that is what became known as a 'Macadamed' road. Writing in 1861 in his books *London Labour and the London Poor*, Henry Mayhew noted that:

> 'Macadamization was not introduced into the *streets* of London until about 25 years ago. Before that, it had been carried to what was accounted a great degree of perfection on many of the principal mail and coach roads. Some 50 miles of the Great North Road, or that between London and Carlisle, were often pointed out as an admirable specimen of road-making on Mac-Adam's principles.'

However he pointed out that the shortcomings of Macadam's roads led to much opposition in parts of the city, particularly those who opposed the macadamisation of Piccadilly.

Opposite above left: The partially overgrown remains of the early eighteenth century pack road at Blackstone Edge in Lancashire – often said to be Roman, but actually laid around 1735.

Opposite above right: The paving of Cooper's Row in Wigan, Lancashire, with smoother lines of flags for carts and carriage wheels to run on, probably dates from the early eighteenth century.

Opposite below: 'Cast Iron Billy', a London omnibus driver. After more than forty years 'on the box' being shaken about throughout each working day on granite-setted streets, Billy suffered from severe arthritis and needed help to get up to his driving position.

Left: A crowded two-horse omnibus running on the recently granite-setted roadway after crossing London's Tower Bridge – a stereoscopic (3D) view from 1895. Tower Bridge had been opened in 1894.

A three-horse omnibus on Hagley Road in Birmingham. Usually a third horse was only added when approaching a steep incline. In York for many years, 'Dobbin' was added to haul trams up Micklegate, then unharnessed and left to make his own way back down the hill where he just stood and waited for the next tram.

Right, below left and below right: Early twentieth century coloured postcards of two-horse omnibuses operated by the London General Omnibus Company. Speeding on 'cobbled' or 'setted' streets in vehicles like these was fraught with danger, with many recorded instances of the driver being thrown to the ground and the coach overturned. Many early horse-drawn tramcars strongly echoed the design of these vehicles.

'The opposition to macadamizing of the latter thoroughfare assumed many forms. Independently of the conflicting statements as to extravagance and economy, it was urged by the opponents, that the dust and dirt of the new style of paving would cause the street to be deserted by the aristocracy.'

In urban areas with heavy traffic, however, even macadamed roads broke up easily under the relentless passage of carriage and cart wheels, not to mention the impact of the hooves of more than 24,000 horses.

Town and city streets in the nineteenth century were therefore usually paved with granite 'setts' – often incorrectly referred to as cobbles – and these created highly durable, if somewhat uneven and noisy, surfaces but with the additional drawback that carriages and omnibuses with poor springing and solid tyres and wheels offered little in the way of passenger comfort.

In Mayhew's day, the actual road surface was referred to as the 'pavement' – appropriate, really, if it had been paved with cobbles or setts – and what we refer to as the pavement today was the 'footpath'. He offered his readers details of comparative costs for the different methods of road construction, noting that 'within the metropolis

proper' there were 40 miles of granite-paved streets, and 1,350 miles of roads with macadamised surfaces.

The advantage of a newly laid macadamised street, he acknowledged, was that it gave passengers a smoother and quieter journey than a paved surface, although nothing like what we expect today.

The drawback, of course, was that the surface was not particularly durable, especially in locations with heavy cart and carriage traffic.

As far as construction costs were concerned, in 1861 macadamised roads cost £44 per mile to lay while paved streets cost £96 per mile. Thereafter, the macadamised surfaces needed constant repair and maintenance. They were dusty in dry weather and muddy in wet weather, and quickly rutted – a busy macadamised road and a lot of horses in wet weather was not a good mixture – whereas the paved streets needed only regular washing and sweeping.

It is remarkable that passenger comfort figured so low in the design of mid-nineteenth century road vehicles, as the first pneumatic tyre had been patented in 1845 by Scots-born Robert William Thomson. It was not widely adopted until fellow Scot, John Boyd Dunlop's tyre was introduced in the 1880s. Dunlop, of course, eventually made a fortune out of it and dramatically improved the comfort of travellers. Solid tyres prevailed on many vehicles until the First World War, however – as is seen by the solid tyres on the horse-drawn omnibuses illustrated opposite.

Fitting a pneumatic tyre to a traditional cartwheel was almost impossible, so the gradual transition from solid to pneumatic tyres required someone, quite literally, to 're-invent the wheel'.

But long before that happened on a large and widespread scale, it would be the passing of the Tramways Act in 1870 which brought the first hints of a smoother journey for those travelling on London's streets.

Above left: In the British Motor Museum in Gaydon, Warwickshire, there is a fascinating exhibit – the Time Road – which recreates the changing character and quality of road surfaces from 1896 to the present day. Even at the dawn of the motor car, 'macadamed' roads could still be found throughout the country, and their surface could deteriorate relatively quickly under heavy use. The motor car seen here standing on the road surface is a 1902 8hp Albion 'Dogcart'.

Above right: A solid-tyred horse-drawn fire engine could wreak considerable damage speeding along a macdamed street. This is a detail from an early Edwardian postcard.

RUNNING ON RAILS

Before dipping into the story of the tramcar, a mention is appropriate here of the origins of the word 'tram'. Originally, 'trams' or 'trums' – a word with its roots probably in Scandanavia – were not vehicles, but the lines of wooden boards on which quarry and mining vehicles were hauled over muddy and boggy ground. They served a similar purpose to the flagstones laid along roadways to smooth the transit of wagon wheels. As such, wooden 'tramways' or 'wagonways' have been found dating back more than 300 years in coal mine workings from the seventeenth century, uncovered during opencast mining in more recent times.

The term was later applied to the primitive trackways laid in coal mines, and eventually to the trucks or barrows which were hauled along those tramways. Indeed, those terms remained in use in coal mining communities in Britain until the end of deep mining.

Over time, those wooden boards were replaced by crude wooden rails, then simple metal rails, and the vehicles which ran over them became 'tram cars' – trams for short.

However widespread the term became it was never as widely adopted in America where a tramway, according to *Webster's Dictionary*, is simply a railway in a mine. The urban systems we think of today as tramways generally became known in America as 'street railways', 'trolleys' or 'cable cars' – the last being a term we usually reserve for the suspended overhead systems which take passengers up mountains at ski resorts. On the ground, however, the distinction between urban railways and tramways is now becoming blurred even in Britain, where twenty-first century hybrid 'tram-trains' – licensed to operate on railway track as well as city centre tramlines and with dual signalling systems to meet the requirements of both – are gaining in popularity.

Passenger tramways in Britain can trace their origins back to the service which was opened between Swansea and Mumbles on 25 March 1807.

Above: A profile section of Loubat-style rail from the tramway in Christchurch, New Zealand. Loubat described this design in his Patent Application No.2789 in 1853. It was the first design which ensured that the rails sat flush with the road surface rather than on top of it.

Below: Alphonse Loubat's 1853 design for a grooved tram rail kept the road surface flat enough for most other users – only cyclists were at risk of getting their wheels caught in the channel. He was granted a provisional patent in Britain on 30 November 1853 – seventeen years before the passing of the Tramways Act – but never submitted a full patent specification. The granite-setted street and tram rails seen here were photographed at the Beamish Museum in Northumberland.

Opposite: A two-horse tramcar on Lord Street in Southport, Lancashire, from a chromo-lithograph published in the late 1890s by the Photochrom Company of Zurich.

Inset: Laying Loubat-style tramlines in Torquay, from a postcard c.1905.

One of the many Edwardian postcards that were produced of the 'Mumbles Train', showing the linked double-decked tramcars. By this period they were hauled by a small side-tank steam locomotive.

Originally known as the Oystermouth Railway, some historians describe it as a tramway, others as a railway, but whichever you choose, it was the first such passenger-carrying transport system on rails in Britain.

The early vehicles were horse-drawn and looked more like stagecoaches on rails than the trams which would become familiar later. But by the Edwardian era, small powerful steam locomotives hauled 'trains' of six or more double-decked tramcars on this hugely popular route. Just as steam had replaced horses in the 1860s, so electric

Horse-drawn trams in Harlesden, from a postcard c.1905. Horse-drawn trams had been introduced in 1888 and electric trams arrived in 1907.

traction with overhead wires would replace steam power in 1929, electric double-decked cars continuing in service until the line closed in 1960.

The first steam locomotive is believed to have been built by Henry Hughes at his Falcon Works in Loughborough, where many steam tram locomotives were built for tramways across the country during the second half of the nineteenth century. It was built to look like a short carriage.

A street tramway, however, was a different proposition to a railway track which was not shared with other users such as horses and carriages. It had to be smooth enough for those other users to run along it without the track itself being an obstacle.

It took some time, however, before the profile of the typical tram rail evolved. The Oystermouth Railway – originally laid to a four-foot gauge – had used L-shaped rails with the wheels running within the 'L'. Early street trams with flanged wheels ran on a track similar to a railway track and, while that worked effectively enough in some situations, it was not appropriate for use on streets shared with other vehicles as the rails sat proud of the roadway. It would be more than forty years after the opening of the Oystermouth tramway before that problem was solved effectively.

The grooved tram rail – also sometimes referred to as a 'girder rail' – was the invention of Frenchman Alphonse Loubat in 1852. He had already been involved in the development of street tramways in New York in the 1820s, but his revolutionary rail design was created after his return to Paris. Set into cobbled streets, it was flush with the road surface and thus did not impede the passage of horse-drawn and, later, motor vehicles.

Loubat filed a provisional patent specification for his rail in London in November 1853 – No.2789 – but for some reason did not submit a full specification within the required time frame and the patent thus lapsed.

Briggate, Leeds, in the 1870s, with heavy horse-drawn traffic sharing the street with horse-drawn tramcars. In conditions like this, the track needed regular cleaning as the channel otherwise quickly filled up with debris, especially in wet weather. In the days of horse trams, the tramway companies employed teams of cleaners to walk the route and clean the rails, removing fallen horse manure and other debris.

The originality of his idea was remarkable, especially as his revolutionary girder rail was designed almost twenty years before the passing of the Tramways Act in 1870 enabled street tramways to be built.

The rail's design has remained fundamentally unchanged in almost 170 years. His one-page provisional specification described the rail:

> 'This invention relates to the application to ordinary roads of a rail having a grooved upper surface, such rail being intended to serve as a tramway for waggons and other vehicles. The rails are set in such a manner in the road as not to project above the surface thereof; they will therefore present no obstacle to the progress of the ordinary traffic. The rail is provided with a longitudinal groove of a U or V shape in section, for the reception of the flanges of the carriage wheels. These improved tram-rails, which may be applied either to macadamized or paved roads, may be bolted in any convenient manner to transverse and longitudinal sleepers, in the ordinary manner of laying down tramways or railways.'

Tramways and railways offered great advantages over horse-drawn coaches – running on smooth rails reduced friction, increased the amount of freight one horse could haul, and in passenger cars gave the occupants a more pleasant ride. It is said that, on rails, a single horse could pull a load ten times greater than on a typical early nineteenth century dirt road.

But in a time before tramways were widespread, others came up with hybrid solutions to moving freight and people around towns and cities. One such proposal, from William Joseph Curtis – who described himself as a civil engineer from Islington – was a complicated, and ultimately unworkable, proposal to engineer a tramcar which could operate both on tramlines and on 'common roads', and he patented it in May 1856 under Patent No.1071. To achieve this, he proposed:

> 'The carriage is made with suitable wheels to run on common roads, but it is preferred that they should be somewhat wider than usual. And in order to cause the carriage wheels when running on tram or rail ways to keep on the rails, there are additional smaller wheels applied to the carriage, suitable for running in or on the rails, and these additional wheels are capable, by levers and connecting rods or suitable apparatus, of being raised and lowered by the driver or other person, hence, when it is desired that the carriage shall be retained on the rails, the additional wheels shall be lowered and become guide wheels for the carriage; but when the carriage is run off the rails on to a common road, the additional wheels are raised and kept up out of the way, and such is the case so long as the carriage is to run on a common road.'

No record has so far been found of such a hybrid vehicle ever being built and put to the test – but the mechanics involved would have been complex and the challenge of aligning the guide wheels with the track when moving from roadway to tramway would have been considerable.

As tramways grew in popularity, a replacement for the horse became increasingly necessary. As the tramcars themselves grew larger, two horses were insufficient to haul them, and three-horse vehicles started to appear. With suitably strong horses costing a considerable sum to buy, an even greater sum to feed and stable, and only a short useful working life before fresh animals were needed, it is not surprising that alternative forms of traction were being explored and adopted by the 1870s.

Frank H. Mason, writing about European tramways in the US-published *Cassier's Magazine* in June 1895 noted that:

'In Germany, good horses for tramway service cost from £40 to £50 each, and their average efficiency does not exceed three years, at the end of which time they are either worn out, or, if salable for breeding purposes or farm work, they bring only from one-fourth to one-half of their original cost. In Dresden, the annual depreciation of street-railway horses from all causes – disease, accident and inevitable wear from hard service in all weathers on hard pavements – is reckoned at from 18 to 22 per cent of their value, and this percentage is said to be still higher in tropical or very cold countries, where only inferior breeds of draft horses are available and the conditions of animal life are less favorable.'

In 'Gas-Motor Street Cars' – which he saw as the future – Mason provided a useful comparison of the costs of using horses against the costs of running a motor tram.

'The cost of keeping a gas-motor car in repair, although not yet fully demonstrated for a long period, is estimated at not more than 5 per cent annually of its original cost, and with ordinary care such a car should last as long as two or three outfits of horses, which latter are moreover, subject to epidemics and to conscription, in case of sudden war, for military purposes. So far, therefore, as experience has yet demonstrated, the mechanical efficiency of the gas-motor car would seem to be assured; and a comparison of its cost of construction and operation with the known expense of working horse, cable, steam and electric tramways in the United States can hardly fail to invest the new motor, as a competitor in the same field, with a serious practical interest.'

A double-decked tram pulled by two horses makes its way along Causewayhead Road on its way to Stirling c.1920. The local tramways continued to use horse-drawn vehicles until 1920. They briefly experimented with a single tramcar powered by an internal combustion petrol engine in the final months before the tramway was closed down.

Laying new tram lines in Christchurch, New Zealand, in 2008. The profile of the grooved rail remains largely unchanged from Loubat's 1852 design.

Above left: Market Street in Manchester c.1906. The tramcars are seen here running very close together, suggesting the photograph was taken at rush hour.

Above right: The horse-drawn Folkestone, Hythe and Sandgate Tramways operated between 1891 and 1921 from Red Lion Square, Hythe to the Sandgate Hill Lift, along the Kent Coast.

Left: The eleven-mile rail line from Carlisle to Port Carlisle was built to carry goods, but demand for a passenger service resulted in horse trams known as 'Dandy Cars' being introduced in 1863, seven years before the Tramways Act of 1870 regulated street tramways. Four cars operated until steam was introduced in April 1914. One of them – later used for years as the clubhouse for a bowls club – is preserved at the National Railway Museum in York.

Tramcars had first operated in London in 1860, using rails which sat proud of the road surface, thus denying access to other users. It was only after an Act of Parliament in 1870 legislated that only trams running on flush rails be authorised to run on streets that the idea gained popularity and tramways were laid in most large towns and cities.

With trams increasing in popularity when Charles Dickens Jr published the second edition of his popular *Dickens's Dictionary of London* in 1888, he listed the three principal operators of tram services in addition to the many horse omnibus companies also serving the city – devoting three pages to the tram timetables and more than ten to the bus companies.

By then 130 miles of tramways had been constructed, and were operated by the North Metropolitan Tramways Company, the London Street Tramways Company, and the London Tramways Company. Dickens listed the routes and frequency of each company's services, together with their fares.

Back then, buses and trams were colour-coded for ease of route identification – when the red London omnibus was introduced that colour-coding was dropped. Most journeys cost one penny, longer routes cost twice that, and the London Tramway Company's routes from Greenwich to Westminster Bridge (every six minutes) or Blackfriars Bridge which ran every ten and a half minutes, cost 3d. Trams from

Below: The Belfast Street Tramways Company started operating horse-drawn tram routes as early as 1872, but planned to replace horses with Hughes steam tractors just five years later – but approval was withheld, and horse-traction continued until 1905.

Bottom: The horse tram interchange in Amsterdam's Dam Square, from a chromolithograph published c.1897 by the Photochrom Company of Zurich. Trams drawn by a single horse remained in service on some routes until the start of the First World War. In London, the last horse tram service ran in 1915.

Opposite top left: Several publishers produced black-bordered 'remembrance cards', lamenting the withdrawal of horses as the traction for tramways. York Tramways Company operated a fleet of ten tramcars and three dozen horses from 1881 until 1909. An alternative version of this card urged 'John Willie' 'to run, to run'.

Opposite top right: While some publishers lamented the passing of York's horse trams, others ridiculed them.

Opposite middle: Leicester operated horse cars from 1874 until 1904 when they were replaced by electric traction.

Opposite bottom: The Nottingham & District Tramways Company Ltd's two-horse car No.32, c.1890. Horse-drawn services were started in the mid 1870s, but by 1880, a double-decked tram hauled by a Hughes steam locomotive unit was being trialled on the network.

Greenwich ran into the city from just after 7am, while the last car to Greenwich left Westminster Bridge just three minutes before midnight. With services operating for seventeen hours a day, several teams of horses would have been needed for each tramcar (and each omnibus).

With four thousand omnibuses and more than one thousand tramcars running in London alone by 1890 – most still horse-drawn – the bus and tramway companies owned and operated more than fifty thousand horses in the city, all requiring teams of workers to feed, stable, groom and operate them.

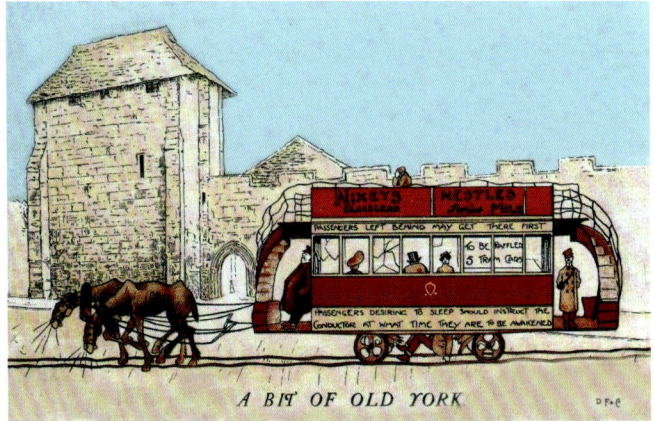

Many companies reckoned that to keep a single two-horse tram operating throughout an average day – given the limitation on how many hours a horse might be expected to haul a heavy vehicle – they required at least ten horses.

So many horses on the street in an ever-expanding city posed an obvious challenge to those whose job it was to keep the streets clean. And there was always the smell. It is little wonder that alternative forms of cleaner and cheaper traction were being explored.

Across the world, experimental systems were being designed, patented and trialled, and by the 1880s, many were ready for public service.

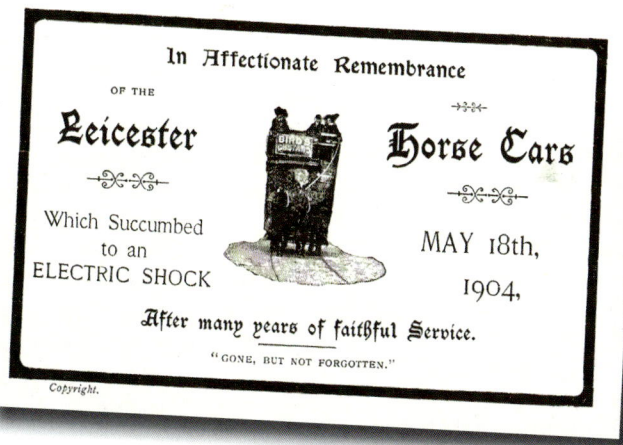

RAISING STEAM

Passenger demand increased exponentially as more and more people migrated to the expanding towns and cities in search of work. As tramcars got bigger to cater for those increased numbers, even a pair of horses was just not up to the task – especially on the many routes where hills were involved. Alternatives had to be sought if heavily laden trams were to travel at more than walking speed.

Many tramways used two-horse trams – and a few, like London's Wood Green to Moorgate route even used three horses abreast – but finding a more efficient and reliable system was essential.

One solution, experimented with in the early 1870s, was to build a steam engine into the tramcar. A four-wheel double-deck design was designed by John Grantham and built by the Oldbury Carriage & Wagon Works. Fitted with a twin-cylinder Merryweather engine, it was briefly tried out in London but not considered a success, thereafter operating for about five years on the Wantage Tramway until 1881.

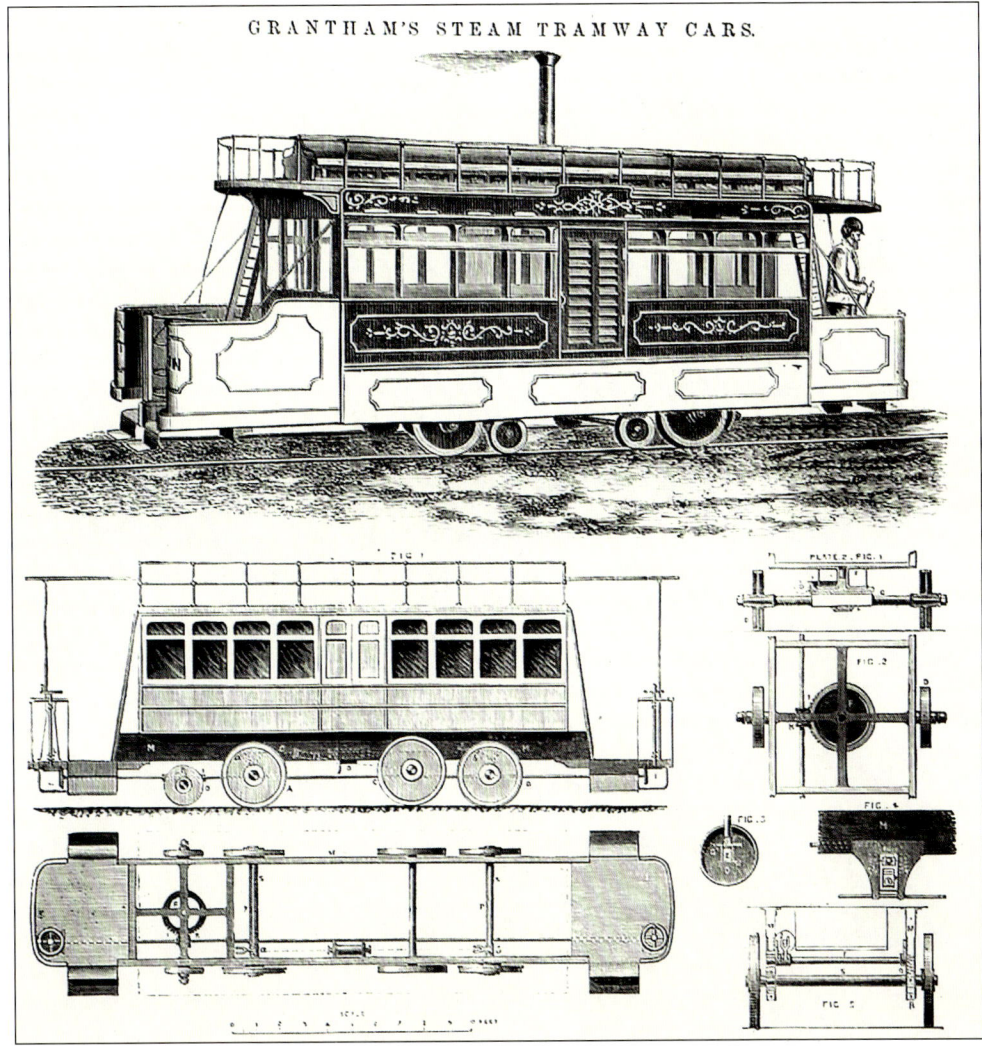

Opposite: Kitson steam tractor Car No.12 taking on water in Market Place, Wigan. The Wigan steam fleet comprised twenty-three steam tractors – nine built by Kitson in Leeds and fourteen built by William Wilkinson at Home House Foundry in Pemberton, Wigan.

Left: This self-contained steam tram was designed by John Grantham and built in 1873 by the Oldbury Railway Carriage & Wagon Works, with a compact Merryweather two-cylinder engine housed beneath the floor. Conditions on the long bench seats on the upper deck near the flue cannot have been particularly comfortable. After its stint in London, the car also ran for five years on the Wantage Tramway.

Thomas Green & Son of Leeds built all the steam engines in Accrington's and Haslingden's fleets, between 1885 and 1898. Accrington withdrew its steam fleet at the end of December 1907, but postcards – published by local stationers, Constantines – showing the steam cars were still being produced well into the town's 'electric' era, some cards, like this one, overprinted as a memorial to the steam fleet.

Right: When Accrington started to electrify its network, some of its steamers were taken over by Rossendale and used until the electrification of that service was completed in 1909.

Below: An advertisement for Thomas Green & Son Limited, illustrating one of their tractors hauling a tramcar from Blackburn Corporation Tramways Company Limited.

For a 'first' it was an ambitious design, seating twenty passengers inside and twenty-four upstairs, and on the Wantage route it was timetabled to operate eight round trips per day.

As originally built, it was said to be underpowered – its boiler, which came from a Merryweather fire engine, was replaced with a much larger one before it was moved to Wantage.

Grantham died just a year after his tramcar first operated in London, and, as his obituary noted:

Wigan photographer Joseph Hulme Aldred took this view of one of the four steam locomotives built in 1883 by William Wilkinson of Pemberton near Wigan and used until c.1930 on the 3-foot gauge Giant's Causeway, Portrush & Bush Valley Railway & Tramway in County Antrim. The tramway was opened in 1883 and part-electrified in 1885.

'The latter part of his life was much occupied in the invention and perfection of a steam tramway car, which has been successfully tried, and for which he held a patent; but as the law of this country prohibits the use of steam carriages on public roads, except, under such restrictions as to render their general employment impracticable, it has only hitherto been worked experimentally.'

A more popular proposition was to replace the horses without replacing the tramcars. So, instead of teams of horses hauling the cars, compact and powerful steam locomotives took over on many of the country's busier tramways.

A major advantage of these powerful little locomotives, which were manufactured by: Kitson & Company of Leeds; Thomas Green & Son, also of Leeds; Merryweather & Sons of Greenwich; Dick, Kerr & Company of Kilmarnock and Preston; and William Wilkinson & Company of Wigan amongst others, was that, in times of high passenger demand, they could easily haul two double-decked tramcars coupled together. However, despite their advantages in efficiency and frequency of service – and generally affordable fares – the noise and the fumes they generated made them unpopular in the confined spaces of city-centre streets.

From the point of view of the tramway companies, however, once the capital cost of the locomotive units had been met, they proved to be no more expensive to maintain and operate than horses.

If any tramways could not meet the capital cost, some manufacturers leased the locomotives. Several of Wigan's Wilkinson-built locomotives, for example, bore plates identifying them as being the property of the manufacturer.

A 'Scott Series' postcard, of the Perry Bar tram terminus, with a locomotive by Kitson of Leeds hauling a trailer built at the Falcon Works in Loughborough. The route had been opened in the early 1880s by the Birmingham Central Tramways Company and remained steam-hauled until 1906.

Right: Paris used steam tram cars fitted with flash boilers designed by Léon Serpollet on several routes from 1895–1913.

Middle: The Wolverton & Stony Stratford tramway, opened in 1886 using three 0-4-0 locomotives built by Hughes & Company of Loughborough – later known as the Falcon Engine and Car Works. A fourth came from Thomas Green & Company and this one from Lokomotivfabrik Krauss of Munich.

Bottom: A Purrey steam tram on the streets of Paris.

Opposite top: On the Wisbech and Upwell Steam Tramway power came from 0-4-0 locomotives designed by Thomas Wordsell and built at Stratford Works by the Great Eastern Railway. The design was the inspiration for Toby the Tram Engine in the Thomas the Tank Engine books.

Opposite bottom: The Wantage Tramway used conventional locomotives as well as skirted tram engines. This little 0-4-0, known as *Jane* was originally built for the Sandy & Potton Light Railway in 1857 but was acquired by Wantage Tramway in 1878, working there until 1925. It is now preserved at the Didcot Railway Centre in Oxfordshire. The tramway also experimented with compressed air traction, but the system proved unsuccessful.

2149. – LES MOYENS DE TRANSPORT A PARIS. – Tramway à vapeur, système V. Purrey
(Cie Générale des Omnibus)

Steam Tram, Wisbech.

When the introduction of gas tramcars was being discussed in the press, a report in *The Belfast Newsletter* on 13 July 1896 suggested that the new trams would be 'appreciated by Lancastrians, who are overrun with smoking and ugly steam trams' – an odd comment, as the rules governing the use of steam locomotives on tramlines were very clear and quite strict.

They had to be relatively quiet, and were not permitted to emit smoke or steam when operating. 'Ugly' they might have been, but they should certainly not have been 'smoking'.

Thus they were fuelled with coke rather than coal, a ready supply of which could be obtained from any one of the country's 1,600 municipal gasworks. Some locomotive designs were fitted with condensers to turn the excess steam back into water and store it in the vehicle's tanks for re-use.

As an alternative to condensing the steam, in 1883 William Wilkinson patented a system of reheating the steam by passing it through superheaters in the firebox – described in British Patents numbers 1631 and 5560. Technically that would not actually eliminate the problem – steam was still being discharged – but raising its temperature did make it invisible.

Having first patented his tram locomotive under Patent No.4268 in 1881, Wilkinson in his Patent No.1631 in March 1883 noted that 'My said improvements are designed to prevent the emission of visible steam or sparks from the funnel and to reduce the noise of the exhaust.'

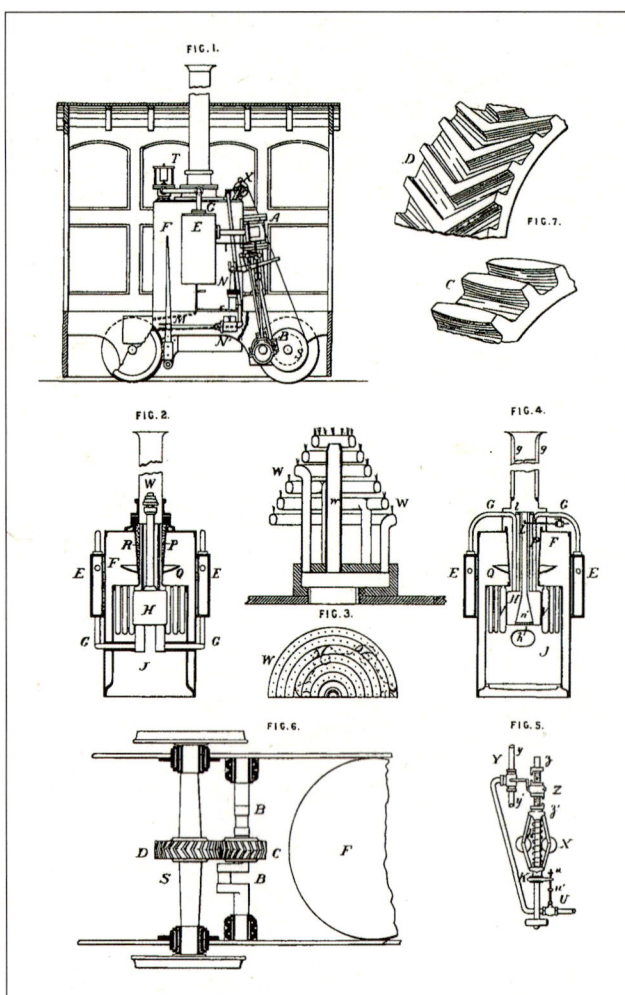
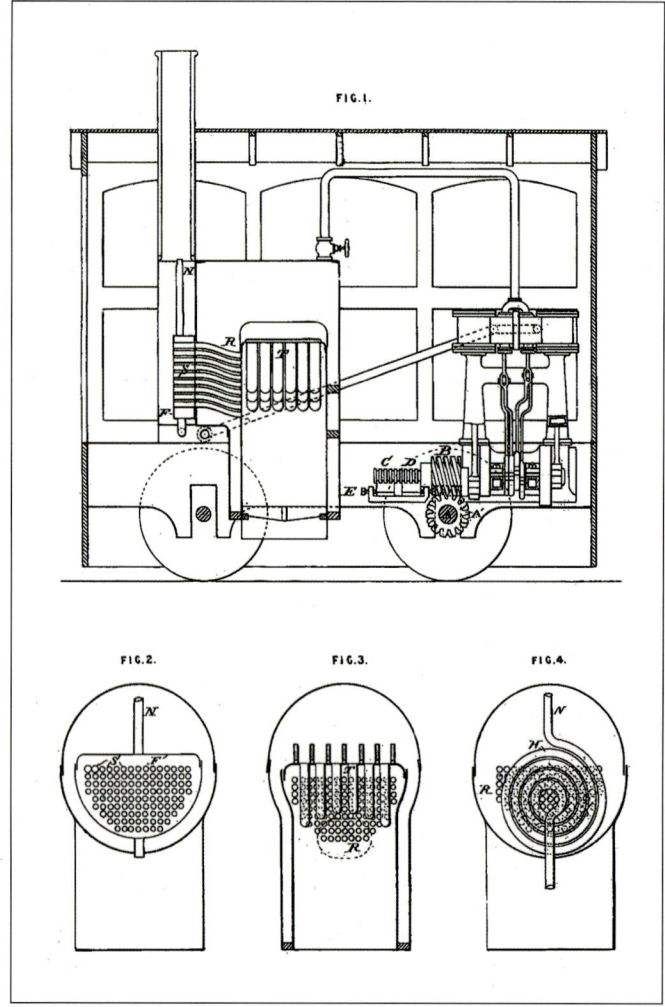

Above left: The layout of the tramcar locomotive as illustrated in William Wilkinson's Patent No.1631, submitted on 31 March 1883 and accepted on 28 September. The patent was described as *Improvements in Traction Engines for Tram, Rail, or Other Roads*.

Above right: Wilkinson's revised design for his superheater was submitted on 28 November 1883 and approved on 28 May 1884. It contained three alternative designs for the superheater tubing.

Just two months after being granted approval for Patent No.1631 *Improvements in Traction Engines for Tram, Rail and Other Roads,* Wilkinson filed another patent application, this time titled *Improvements in Traction Engines for Tram and Other Roads*.

In this application, he offered three different designs for the superheater, claiming that each of them enabled the engine to operate 'without visible or audible emission of steam'.

However counter-intuitive it might have been, the success of the system led other builders to license the rights from Wilkinson – amongst them Beyer-Peacock of Gorton and Thomas Green of Leeds. An example of a 'Wilkinson's Patent' engine built by Beyer-Peacock is preserved at the Tramways Museum in Crich.

As to the requirement that they operate quietly, contemporary reports suggest that, although they were quieter, such a state of affairs was never actually achieved.

For the safety of pedestrians and animals, the machinery on tram locomotives operating in Britain had to be completely enclosed and even the wheels had to be concealed behind skirts extending down from the bodywork to just above the level of the roadway. It is clear from contemporary photographs, however, that such strict safety rules were apparently not always observed on steam trams in mainland Europe.

In Germany and France, where many steam tramcars were self-contained units with the engine in the front of the vehicle, skirting was usually only fitted around the 'hot bits'.

The introduction of the steam tram brought about a major change in the required skills of the tramway's workforce. There was no longer the need for stable hands, farriers and the many other occupations associated with the working and welfare of large numbers of horses – and more obviously, there was no need for most of the tens of thousands of horses.

Instead, there was a sudden demand for technically trained drivers, firemen and workshop staff equipped with the wide range of engineering skills necessary to maintain and operate the steam engines. The driver was now also an engineer.

Where previously a tram driver had only been required to understand the means of controlling his horses, and keep them fed and watered, he now needed to understand the intricacies of a small but powerful steam engine, keeping water levels and steam pressure under his constant watch – in other words, doing everything a railway locomotive driver had to do, but on a smaller scale and within a much more confined space – which he shared with his fireman and their supply of coke.

Despite all these new challenges, there were remarkably few serious accidents in the many towns which adopted the steam tram.

There were also a few key procedural changes involved in operating a steam tramway. With a horse tram, all that was required at each terminus was to unharness the horses from one end of the car and walk them to the other end – and even that was not needed on one Manchester route – the Manchester Carriage and Tramways Company used Eades Patent Reversible cars, using the horse to rotate the tram body on its underframe, avoiding the need to uncouple and recouple it.

With a steam tram locomotive, however, a turning triangle was needed if the vehicle only had a driving position at one end, or a passing loop if it had driving controls at both ends.

In urban areas, the steam tram had all but disappeared before the First World War, largely replaced by electric traction, and with the benefit of hindsight, it proved to be something of a cul-de-sac in tramcar development.

The Wantage Tramway was the first in Britain to use steam, using John Grantham's steam tram from 1876 to 1881 (see page 27). Engine No.6 was built by Fox Walker & Co of Bristol in 1882. It used a design patented in 1879 by James Matthews of Bristol and was sold to Wantage in 1885 after the tramway's failed experiment with compressed air (see next chapter).

A steam-hauled tram in Birmingham's Old Square, from a postcard mailed just a few months before steam trams were withdrawn in the city.

One of Merryweather & Sons' steam tractor units being trialled on the North London Tramway Company's lines. The company operated a fleet of steam tractors, some built by Merryweather and the others by Dick, Kerr & Company. The large, long wheelbase tramcars were built by the Falcon Engine and Car Works in Loughborough, formerly Henry Hughes & Company. Merryweather & Sons could trace their history back to the middle of the seventeenth century and, by the 1880s were building fire appliances, steam pumps and steam tram locomotives at their Tram Locomotive Works in Greenwich. They also built locomotives for the North Staffordshire Tramway and the Alford & Sutton Tramway in Lincolnshire, as well as many tramways in mainland Europe. The original glass negative of this 1880s image survives in the collection of the Australian Railway History Society.

Despite having been introduced as a means of saving time and money, on many tramways the steam locomotive actually had turned out to involve a greater initial capital outlay and higher maintenance costs than the horses it replaced.

The heavy locomotives inflicted considerably more wear on the tracks than the lighter horse-drawn vehicles that they replaced, requiring the lines had to be re-laid with a much a more substantial gauge of rail. Even so, the wear on the tracks was considerable.

A 'memorial' postcard marking the end of Birmingham's steam trams, complete with a specially composed poem. Several of the criticisms of the steam trams contained within the verses would later be targeted at gas tramcars as well.

The passing of Birmingham's steam trams was certainly not universally applauded, and a number of commemorative items were produced to mark the occasion. This postcard shows a tram at the Perry Bar terminus, but the same image was used on a number of other cards.

When proposals for gas traction were being put forward a few years after steam had been introduced, some detailed infrastructure costings were carried out which showed that when fuel and maintenance were taken into consideration, the additional costs were much more significant than had originally been predicted.

Operating a cumbersome steam locomotive on a street railway was also potentially dangerous, its smoke unpleasant enough for it to be banned in many towns and cities. A better system was urgently needed.

Cromwell House, Highgate.

PRINCES St BOTANIC GARDENS INVERLEITH ROW AND GOLDEN ACRE.

EDINBURGH AND DISTRICT TRAMWAYS COY LTD

BATTERIES, CONDUITS, CABLES, OIL AND COMPRESSED AIR

Tramways planning to replace their horses had several available alternatives to the steam-hauled car. Each had its advantages and each its drawbacks, some requiring the installation of expensive infrastructure along the entire planned route, others not. Discussing the options, the journal *The Engineer* published details of a novel solution in its issue for 21 March 1890.

> 'Electrical propulsion is theoretically more perfect, inasmuch as the generator has a higher efficiency… and an electrical motor a higher efficiency than an air motor; but it has to be admitted that these efficiencies have not hitherto enabled electrical engineers to propel tramcars with that economy which is necessary to constitute a sufficient reason to adopt electricity in place of other means of propulsion.'

The system discussed in *The Engineer* had been developed and patented by John Hughes of Chester in 1888 and developed by Hughes & Lancaster of Wrexham, Denbighshire. It ran on compressed air and Hughes clearly had great faith in his invention for in the four years after the granting of his British Patent, he patented it in France, Belgium, Germany, Italy, Spain, the Austro-Hungarian Empire and the United States of America. In the course of the research for this book, however, the only evidence found of it being used anywhere was in Germany, its limited appeal most likely because of the high infrastructure costs involved in setting up the network of pipes.

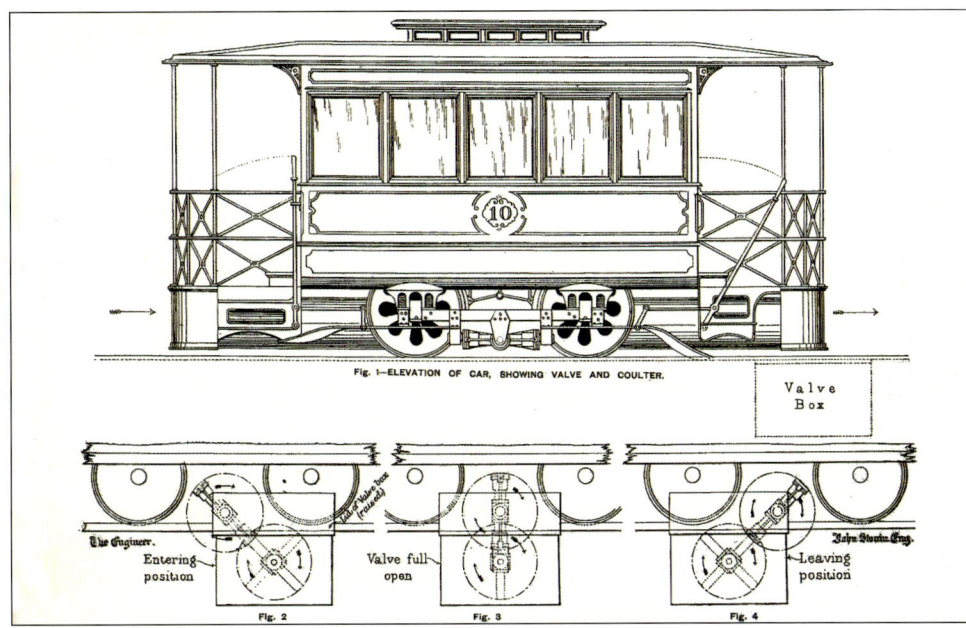

Opposite above: Two of the Highgate Cable Tramway's open-topped double-decked cars, seen here in a postcard published by Charles Martin of 39 Aldermanbury, London. As with so many double-decked cars of the period, the upstairs bench seating ran the length of the tramcar, passengers facing out on both sides.

Opposite below: The Edinburgh & District Tramways Company was set up in 1893 as a subsidiary of tramcar-builder Dick, Kerr & Company of Kilmarnock and Preston. This photograph of the company's first cable tram, a double-bogied open-topped car, was taken by Thomas Polson Lugton whose studio was at 14 East Mayfield in Newington, Edinburgh.

Left: The 21 March 1890 issue of *The Engineer* featured Hughes & Lancaster's compressed air tram system.

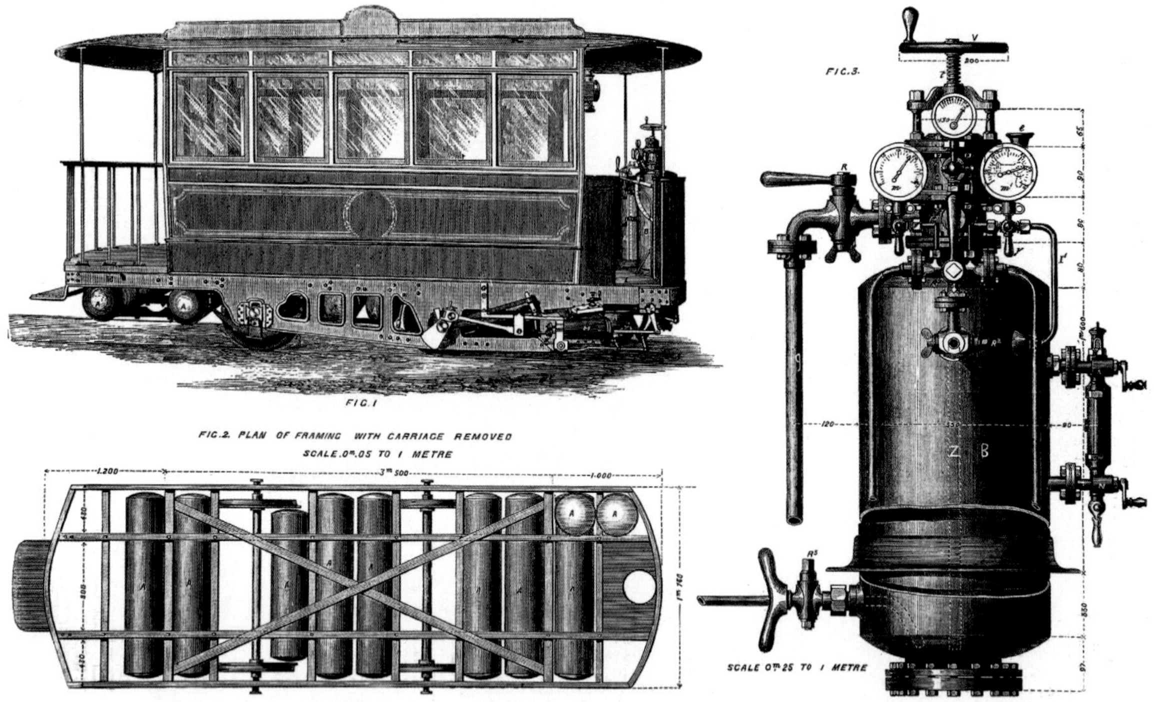

BATTERIES, CONDUITS, CABLES, OIL AND COMPRESSED AIR • 39

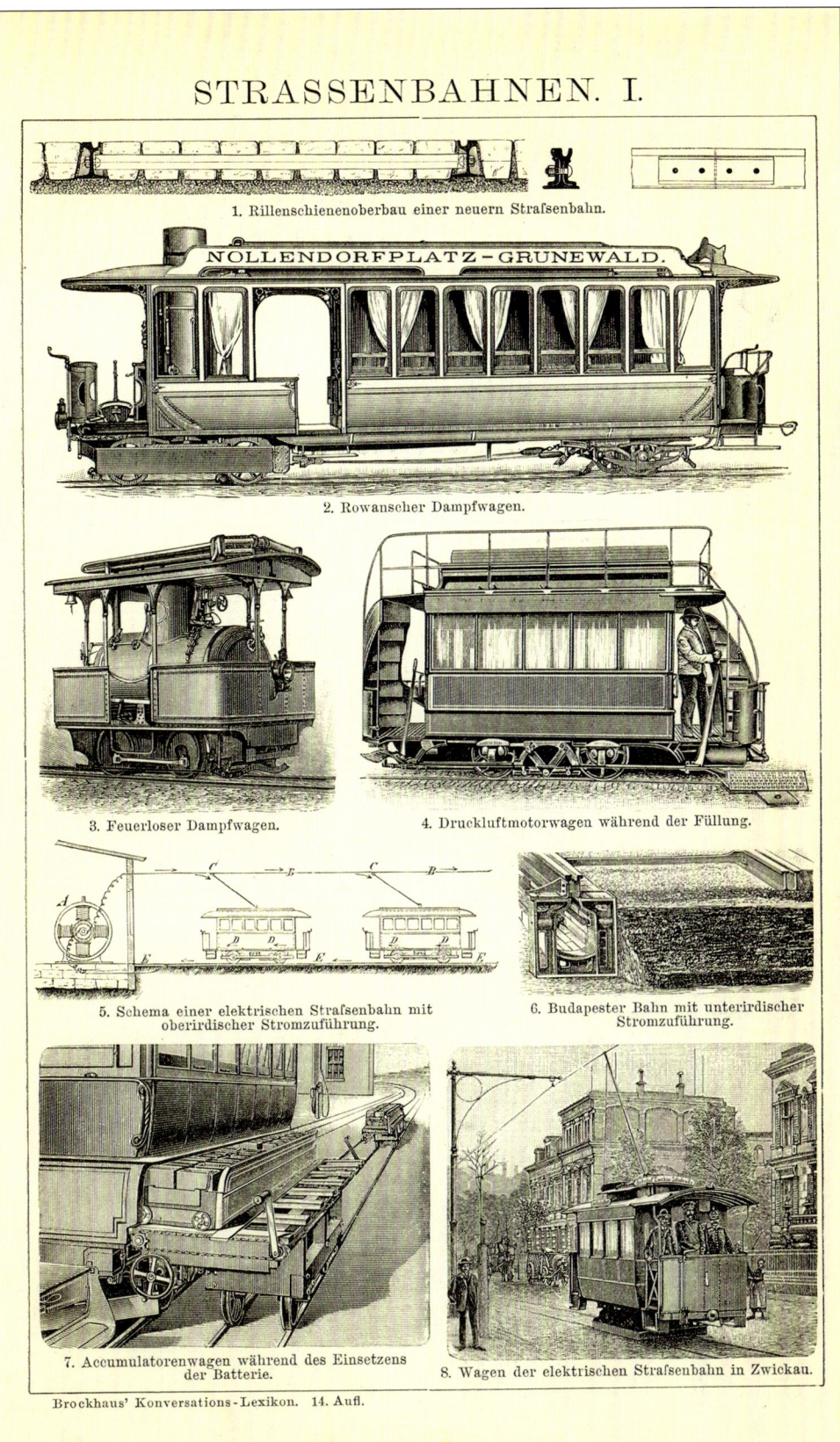

Left: From the 14th Edition of *Brockhaus' Konversations-Lexikon*, published in 1896, an illustration of several of the many types of tram then in use in Germany and Hungary. The drive mechanism for the gas tram which was already operating in Dresden and Dessau appears on the following page. From the top: 1. The grooved structure of a new tram rail; 2. The Rowanscher Steamcar; 3. A Fireless Steamcar; 4. A Compressed Air Motorcar being recharged with compressed air; 5. A schematic showing the now-common overhead catenary used for Electric Trams 'with above-ground power supply'; 6. The Budapest tramway used an underground power supply in a central channel between the rails; 7. On a battery tramcar, the accumulator trolley was used to support the battery rack when discharged batteries were being replaced with freshly charged ones; 8. An electric tramcar on the streets of Zwickau in Saxony, central Germany.

Opposite: A detailed description of the Mékarski compressed air system was published in *The Engineer* on 4 March 1881 together with annotated illustrations of its general layout and control systems (*top*). Note the line of cast-iron compressed air tanks visible beneath the rear of the car. A slightly different version of the car, known in France as an 'automotrice', was published in the French journal *Illustration* on 20 November 1875.

40 • THE GAS TRAMCAR

It was not the first to harness the potential power of compressed air, nor was it the simplest, but it was certainly ingenious, albeit potentially fraught with problems. In his patent specification Hughes made much of the fact that the system worked 'automatically':

'At certain intervals along the "air-main" we provide devices provided with valves which for the convenience of description we hereinafter call "supply" devices. Upon the car or in connection with the receivers, which are attached to or form a part thereof, we also provide devices provided with valves which we call "receiving" devices. These devices are so constructed that as the car continues running the receiving devices may be cause at the will of the driver to engage with the supply devices and so form a temporary connection between the air main and the receiver, disengaging themselves immediately afterwards as the car moves forward.'

Key to the system was a means of delivering a controlled supply of compressed air to the tramcar at regular intervals and that, of course, required a network of high-pressure air pipes to be laid, with supply points installed along each route, all fed from a central pumping station. The supply points could be up to a mile apart on level track, but very much closer together if the tramcar had to ascend even moderate gradients.

Tramcars fitted with much simpler pneumatic motor systems were already in use in several cities in France and Switzerland, but central to Hughes' motivation for perfecting his system were the twin concerns of weight and safety.

Existing pneumatic systems involved the cars carrying their own supply of compressed air – in heavy-duty cast iron cylinders – thus, he asserted, reducing the efficiency of the car.

Above: Cross-section of the Mékarski tram's pre-heating chamber with the controls and regulator valve above. The first version of his tramcar required a small coal-fired water heater to stop the cylinders freezing as the air decompressed. As patented in 1875 and 1876, he offered a second option of a jacket round the decompression chamber filled with pre-heated water which would be replenished as the air tanks were refilled.

Right: Over the years, the Mékarski motors became much more powerful – and thus much more versatile and able to power double-decked trams hauling a trailer. This postcard was published in Paris in 1910 showing the effects of one of the River Seine's regular winter 'inundations'.

Les Inondations de Paris en 1910
Tramway sur les Quais

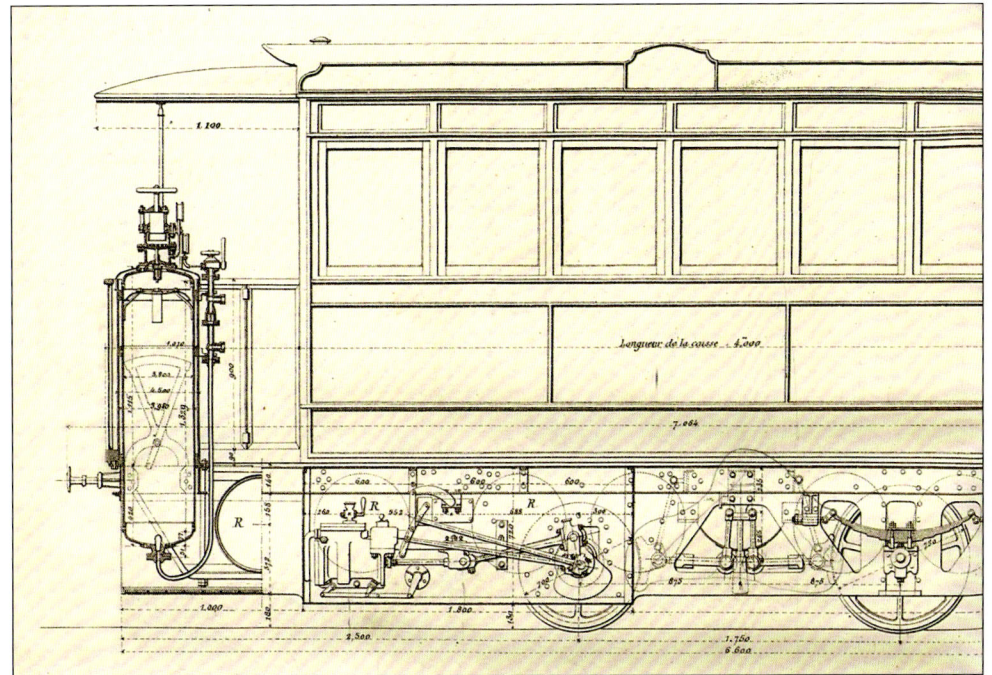

The drivetrain of the Mékarski tramcar, as illustrated in the journal *Le Génie civil: revue générale des industries françaises et étrangères* published in Paris in October 1884.

The same argument had been voiced against the heavy steam locomotives hauling former horse cars on many tramways – the increased weight of fuel meant either fewer passengers or slower trams.

Hughes' other concern was safety, clearly fearing that a cylinder might fracture and explode – the already-established system which had been designed by Paris-based engineer Louis Mékarski in the 1870s used tramcars which were, indeed, much heavier than Hughes' alternative, and they carried their supply of compressed air in high-pressure tanks beneath the floor.

Despite the concerns expressed about the safety of the compressed air tanks, Mékarski tramcars were produced in their hundreds and widely and very successfully used in several cities in France – including Paris, Nantes and La Rochelle – as well as in Bern, Switzerland.

The first Mékarski cars were introduced in Paris in 1876, and were modified and improved over the years. The system was introduced in Nantes in 1879 and by the end of the century, their fleet numbered more than ninety cars. The compressed air system continued in use until towards the end of the First World War when replaced by electric traction.

Mechanically, it was a complex system, as compressed air cools as it is decompressed, and this caused frost to build up in the cylinders. A remedy involved a small steam engine to pre-heat the compressed air before it reached the cylinder, and that meant that a small coal- or oil-burning heater had to be included in the driving cab.

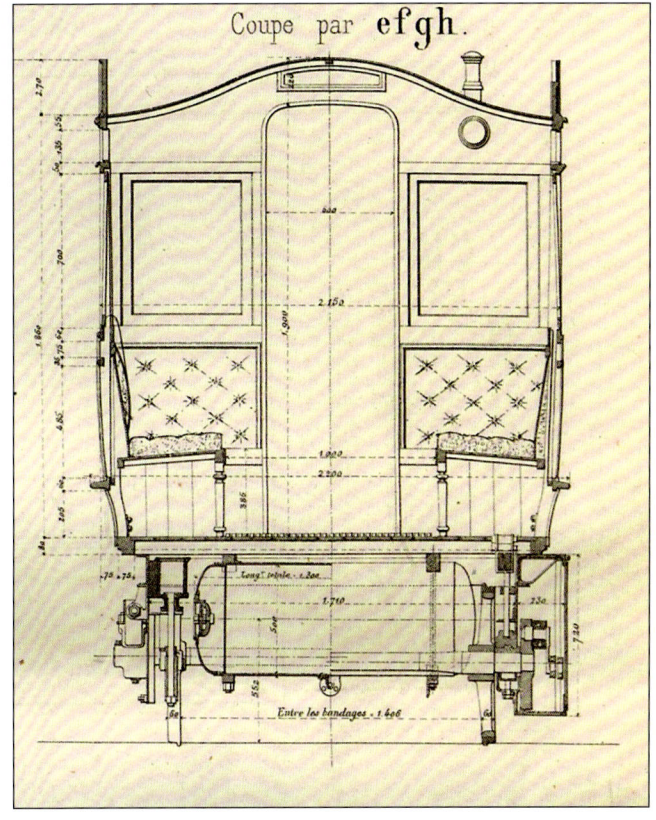

Also from *Le Génie civil*, the end profile of the tramcar showing the compressed air tanks beneath the floor.

Right: Two of the Mékarski trams in operation in La Rochelle, c.1905. The system had been introduced in 1901 to partially replace steam traction. Mékarski trams were also introduced in Paris in 1876. In Nantes they ran from 1879 until 1917, Vichy 1895–1927, Aix-les-Bains 1897–1908, Le Mans 1879–1917. The Saint-Quentin system commenced operations in 1899.

Below: A Mékarski car on the Berner Tramway in Bern, Switzerland. The compressed air system was introduced in 1890. The line ran from the depot at Bärengraben to Bern railway station and on to the Bremgartenfriedhof cemetery. As Mékarski cars were usually single-ended, turntables were necessary at both ends of the line. When Bern was planning its second tramway, the company reverted to steam traction as the compressed air cars could not cope with the route's undulating topography.

While critics made much of the occasions when the tramcars ran out of air before they reached a compression station for a refill, there is little doubt that these were powerful and efficient machines – their power is evidenced by the numerous photographs which survive of them coupled to powerless trailer cars.

What John Hughes might have thought about the safety implications of later gas-motor trams carrying their highly inflammable gas supply in simple rubber bags stowed beneath the seats, we will never know.

With compression technology now much more advanced, the age of the compressed air tram might be about to return – that possibility is discussed in the final chapter of this book.

BATTERIES, CONDUITS, CABLES, OIL AND COMPRESSED AIR • 43

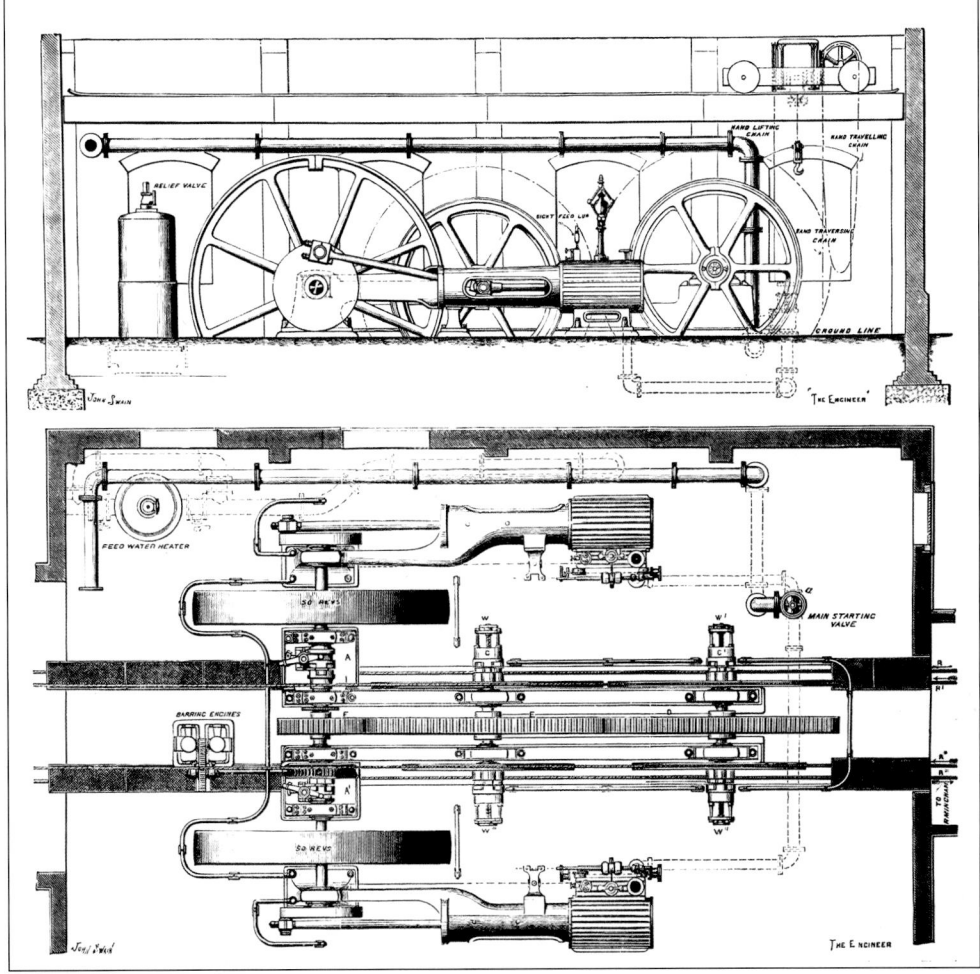

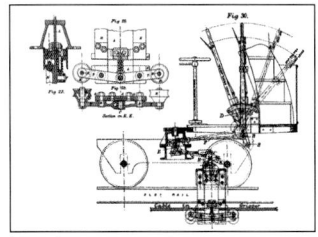

Above: Also from *The Engineer*, the mechanical linkages between the tramcar and the cable in its conduit in the roadway.

Left: From *The Engineer*, 29 June 1888. Two Tangye steam engines powered the cars on Birmingham Cable Tramway's 1.25 mile track. Each 3⅜ inches (86mm) diameter cable – wound from 114 wires – hauled trams built by the Falcon Works at Loughborough.

A hybrid gas/pneumatic system was designed and patented by Howard Lane, a Birmingham engineer in 1896 – a full patent, No.12,901 *Improvements in Machinery or Apparatus for the Propulsion of Tram-cars and Other Vehicles,* being granted on 11 June 1897.

He proposed using a gas or oil engine as the primary motor and a hydraulic motor as the secondary source. As the gas motor ran continuously whether the tram was in motion or not, he proposed using its energy to charge a pneumatic accumulator when the tram was stationary, describing it thus:

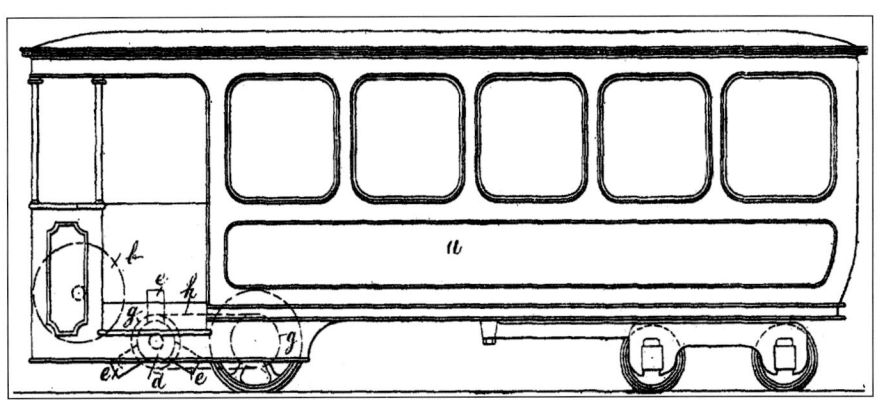

Below: Howard Lane's 1896 six-wheeled single-ended hybrid gas/hydraulic tramcar, intended for one-man operation. On this drawing, the gas or oil engine, marked 'b' with its large flywheel, sits at the front with the hydraulic motor behind it, sitting just ahead of the front wheels.

'In order to put my invention in operation I arrange upon the tramcar a gas or oil engine of a suitable size and character. This motor runs continuously, independently of the stopping or starting of the vehicle. In the case of a gas engine, the gas is stored in steel cylinders into which it is charged at the terminal stations once or twice a day. This engine or motor continually pumps oil

An open-topped cable tram on Brixton Road, from a postcard c.1904. When the line was first opened, passengers travelling from Westminster Bridge to Brixton Hill would start off being drawn by two horses, with those being replaced at Kennington by the 'gripper car' described by Hiram Maxim. By the mid 1890s, the gripper mechanism was housed within the tramcar itself.

Right: An electric tram to Shoreditch c.1922, running along Kingsland High Street. This part of the electric conduit system had originally been built with an outside live rail but was changed to a central conduit pick-up in 1921.

Opposite: Converting the tramway in Brixton to electric conduit pick-up. The London County Council introduced electric trams with conduit pick-up in 1903, replacing both horse-drawn and cable-hauled cars. A small number of lines operated both conduit and cable for a time – a complex infrastructure challenge. The laying of the conduit system on the route along Gresham Road, Brixton Road and Stockwell Road in Brixton in 1907 caused major disruption, but as can be seen from the trackwork being laid, and engineering challenges were considerable. The electric trams were relatively cheap to run, and parts of the system lasted into the 1950s. It was, however, more prone to breakdowns than overhead catenary, and much more difficult to repair when things did go wrong.

or other non-elastic fluid into a pneumatic or spring accumulator. The car itself is driven by a hydraulic secondary motor which is controlled by the attendant's lever, which is so arranged that it reduces as required the stroke of the hydraulic motor, and converts it when desired from a motor to a pump, by well-known methods of valve arrangement. This secondary motor is driven by the liquid from the pneumatic accumulator first described, but when it is converted by reversal of its valves into a pump, then it supplements the gas motor, by forcing the liquid into the accumulator. In this way the momentum of the tram-car is usefully stored when stopping or going down hill, and the power gained can be re-used.'

Lane suggested that a compact and lightweight two- or three-cylinder gas engine 'made of light forged steel' could operate the system, resulting in a much lighter

BATTERIES, CONDUITS, CABLES, OIL AND COMPRESSED AIR • 45

tramcar than those steam-hauled vehicles operating at the time. Its benefits, he claimed, were clear:

> There is, of course, an absence of steam, noise and smoke, stoking and engine driving. The motor is placed on the tram-car itself, the vehicle being complete and capable of being managed by one man, there is economy in fuel and repairs, small capital expenditure, the power usually lost in stopping and in brake action is saved and utilized in imparting motion to the car.'

He also cited reduced wear and tear on the track due to the lightness of the vehicle, and although not mentioned, the single-ended vehicle could utilise the turning triangles at

Electric conduit trams crossing Westminster Bridge in London c.1910. The initial order for 100 tramcars was placed with Dick, Kerr & Company of Preston – who built the 30-horsepower motors – with bodies by the Electric Railway and Tramway Carriage Co. Ltd. One bogie in each pair was modified to take the plough for the conduit current collection used by the London County Council.

An open-topped cable tram turning from North Bridge on to Princes Street, Edinburgh. A short length of the track, complete with cable slot, has been preserved in Waterloo Place, now incorporated into a traffic island. It was formerly part of the track which ran in front of the Post Office and Register House and acted as a terminus/interchange.

termini already in place for the steam trams. It was a clever idea, but in the absence of evidence of successful trials it remains no more than that.

If the cost of installing the pipework for the Hughes compressed air system was considered excessive, it is all the more remarkable that cable-hauled trams became so widely used, as the cost – and the disruption to other traffic using the streets while installing their conduit – was huge And yet, cable-hauled systems using large steam engines in bespoke engine houses were adopted by several towns and cities – London, Birmingham and Edinburgh all developing networks. Systems were also installed in Matlock and up the Great Orme in Llandudno, North Wales.

Of the majority of the 130 miles of tramways in London at the time, *Baedeker's Guide to London* noted that 'Horses are still the main motive power.' But a novel system had been adopted for the Highgate Cable Tramway, opened in 1884, and its workings were afforded a brief explanation.

> 'The Highgate Cable Tramway, the first of its kind in Europe, opened in 1884, ascends *Highgate Hill* from the *Archway Tavern*; the cars start every 5 min. (fare 1d.). The motive power is supplied by an endless wire rope placed in a tube below the surface of the road and kept in motion by a stationary engine. Connection between the car and the rope is effected by means of a 'gripping attachment', passing through a slit in the middle of the track. The rope runs between the jaws of the 'gripper', which the driver closes when he wishes to start the car, reversing the operation and applying the brakes when he wishes to stop.'

London was, indeed, the first to run a cable tram service in Britain – Liverpool had investigated such a system the previous year but decided against the infrastructure costs, which would have been considerable. Birmingham introduced cable trams three years after Highgate, followed the next year by Edinburgh in 1888.

The Brixton cable system opened in 1891 and drew immediate criticism from none other than Hiram B. Maxim – inventor of the Maxim machine gun and the 'Maxim's Flying Machines' which still grace Blackpool's Pleasure beach today. In an article in the *Engineering Magazine* he poured scorn on the over-elaborate system on the line, writing:

> 'A short time ago an American cable line was established in Brixton, a suburb of London. Upon first visiting Brixton I failed completely to recognize the system, as each car was provided with a small and extremely ugly locomotive. Upon closer inspection, however, I found that the locomotive carried simply the clamping device. Upon asking the 'driver', or the man at the clamp, the object of the apparatus, he said:
>
> 'Oh, this is the locomotive. It draws the car.'
>
> 'Oh,' I said, 'how nice! Please explain it.'
>
> 'Well, underground here is a wire rope; this 'ere thing goes down through this 'ere slot and clamps the rope, and the rope pulls the locomotive, and the locomotive pulls the carriage, don't you see?'
>
> 'What is the object of the locomotive?'
>
> 'Why, to draw the car, of course.'
>
> 'But why not put the clamp on the car and dispense with the locomotive altogether?'
>
> After he had thought the matter over a short time, I asked again: 'What is the use of the locomotive?'
>
> His reply was: 'I'll be hanged if I know.'
>
> Now, if this system had been introduced into a country like Germany, France, or Spain the natives would have had sufficient respect and

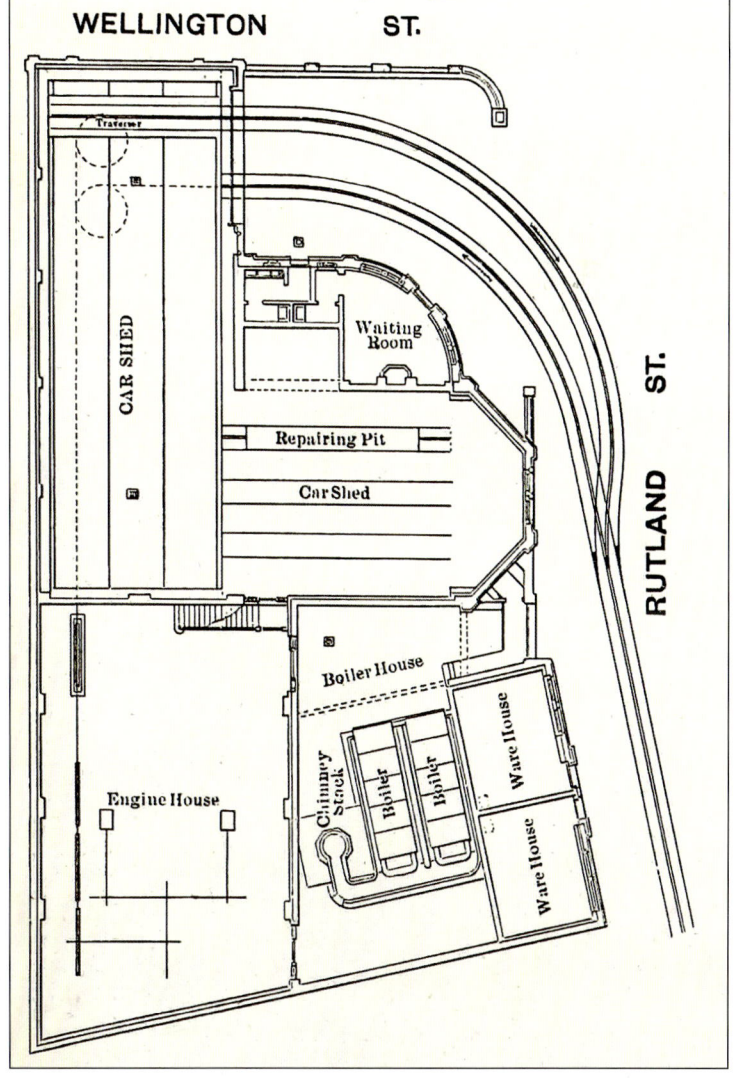

The plans for the cable-tram Engine House and Workshop built in 1891 by the Matlock Cable Tramway in Derbyshire. It was the steepest tramway on public roads in the world, the system having to negotiate an 18 per cent gradient along Bank Road. There was only a single track with a passing loop at midway. The tramway and its shed were described in *The Street Railway Review* in 1891, the service opening in 1893. The line ceased to operate in 1927. There were three double-decked tramcars in the fleet, built by Dick, Kerr & Company in Preston.

48 • THE GAS TRAMCAR

Car No.118 making its way along Portobello High Street, Edinburgh. The cable channel between the rails can be seen. Edinburgh & District Tramways ran cable trams out as far as Joppa from January 1893 until 1923. The cables were driven from an engine house in Portobello. The line ran from Waterloo Place at the end of Princes Street *(see previous page)* through Portobello to Joppa where passengers had to change to Musselburgh's electric trams.

Edinburgh cable trams No.100 and No.169 at their Joppa terminus c.1915.

An early postcard of the Great Orme Tramway. The up and down lines share a common rail for part of the way. From Llandudno's Victoria Station to Halfway Station, the cable is carried in a conduit in the street. The upper section crosses open fells, with the cable run between the rails.

confidence in American engineers and systems to have put it up in the exact manner that it was imported; but as the English engineers were used to a locomotive and wished to make some change in the American model, they added the 'locomotive,' which certainly looks very awkward, and is, without question, superfluous.'

Setting up such a cable tramway required heavy investment in building and equipping engine houses, and major road works to lay the channels. Such systems operated best when the cable ran in a straight line from the engine house. Where there were corners the installation was more complex in order to limit drag on the cable which severely reduced efficiency.

Operating expenditure was also high with staff at the 'power station' to operate and maintain the engines, and maintenance crews working along the route to keep the channel clear and periodically replace worn cables and cable wheels.

The cars, however, ran quietly and emitted no smoke or fumes on the streets, the only emissions coming from the chimney of the engine house. So, a cable-hauled system was pollution-free – unless, that is, you lived downwind of the engine house.

The Edinburgh Northern Tramways Company Limited opened its first cable line from the city centre north to Trinity in January 1888 along a route which included steep hills and some challenging curves – the engine house at Henderson Row with its massive steam engine and multi-cable drum and flywheel drove the six miles of cables and up to fourteen tramcars at any time was not, as might be expected, at the highest point of the route, but well down the hill towards Canonmills.

The route operated by Edinburgh and District Tramways ran from Waterloo Place along the Firth of Forth as far as Joppa, where passengers wishing to continue along the coast had to change to Musselburgh's electric trams, until the Edinburgh section of the track was electrified and the two sections were connected in 1923.

A fragment of the facade of the Henderson Row tram shed and engine house has been incorporated into the 1991 offices built for the Scottish Life insurance company.

Two pulley wheels in the remaining facade of the former Henderson Row Engine House are all that remains of the Edinburgh Northern Tramway's cable system. The Engine House and Depot were the work of the leading Scottish architect William Hamilton Beattie (1842–98) who also designed the Jenner's department store on Princes Street and the North British Hotel – now the Balmoral Hotel – between Princes Street and Waverley Station

'Le tramway funiculaire de Belleville' was a two kilometre cable-hauled tramway which ran up a hill in eastern Paris between the Place de la République and the church of St. John the Baptist in Belleville. It was opened in 1891 and operated until 1924.

Of course, there were drawbacks. If a cable broke – and they did periodically – the entire line was out of action until it could be replaced, and there were reports of some horrific accidents with cyclists getting their wheels caught in the cable channel and being thrown off their bikes.

The last town to introduce the system in Britain – Matlock in Derbyshire – like Edinburgh had some pretty steep hills which meant the cable system was the only realistic option until the advent of more powerful compact electric motors. Their first cars ran in 1893, their last in 1927, four years after Edinburgh went fully electric.

To the uninitiated, the trackwork for electric traction using a conduit pick-up looked similar to the cable system in that there was a slot in the centre of the track. But whereas the cable system used a gripper lowered into the slot to grip the moving cable, the conduit contained the power supply conductor rails at the bottom of the channel.

A T-shaped probe – known usually as a 'plough' – was lowered from the tramcar into the conduit where it made contact with the live and return conductor rails on either side of the channel, completing an electrical circuit.

Like the cable system, the infrastructure costs for conduit pick-up were high, but for those towns and cities where there was a strong reluctance to put up overhead catenary – London being an important example – it offered a viable, cleaner alternative to horse or steam traction.

The system achieved greatest popularity in New York, while in Britain its use was limited to certain areas of London and part of the Blackpool seafront. In that latter location, the system was relatively short-lived, as sand kept blowing into the conduit and making electrical contact somewhat erratic.

Today the only cable-hauled tramway still operating in Britain is the seasonal line from Llandudno to the top of Great Orme in North Wales, still using its original cars built in 1902 when the line first opened. The Halfway Station steam winches were replaced by electricity in 1957.

One of Blackpool's electric conduit tramcars from 1885 is now preserved at Crich Tramway Village, but as there is no conduit system in existence there, on the occasions when it is run it draws its power from a battery pack.

BATTERIES, CONDUITS, CABLES, OIL AND COMPRESSED AIR • 51

Carrying an array of wet cell batteries was a system pioneered in Germany more than a century and a quarter ago *(see illustration page 39)*, but which found only very limitied favour in Britain.

In 1883 the West Metropolitan Tramways Company experimented with a battery-powered car designed by Anthony Reckenzaun (1850–1893), a pioneer in the development of wet cell accumulators. He is believed to have supervised the design and construction of the tramcar itself.

It ran between Acton and Kew Bridge in London, the batteries manufactured by the Electrical Power Storage Company of Great Winchester Street and West Ferry Road, London for which company Reckenzaun worked.

Reckenzaun left the English Power Storage Company two years later to set up his own company, and his improved system was trialled on the South London Tramways Company's lines in 1885, but the considerable weight of the batteries meant the tram was very heavy and seriously under-powered.

It was trialled on several tramways, but while there seems to have been little interest in Reckenzaun's system in Britain, it achieved greater popularity in the United States when it was licensed to the Electric Car Company of America.

Modern battery technology might very well give battery trams a future, just as it is doing with cars and commercial vehicles.

This illustration of Anthony Reckenzaun's battery tramcar was published in 1883 in *The Graphic*, a weekly illustrated newspaper published by William Luson Thomas's company, Illustrated Newspapers Limited.

1. The Car.—2. The Interior of the Car, With One of the Cushions Removed to Show the Accumulators.—3. The Starting and Reversing Handles.
THE NEW ELECTRIC TRAMCAR AT KEW BRIDGE

THE GAS REVOLUTION

In the late nineteenth century, there were more than 1,600 gasworks operating around Britain. Some were small, supplying very local needs, while others were massive, feeding gas into the evolving grid which supplied big cities.

Only three 'town gas' works survive – one each in Northern Ireland, Scotland and England – all of them small, and all three of which have been preserved as if the workforce has just clocked off at the end of a shift.

Few innovations were more influential in changing the quality of life than the introduction of coal gas for lighting in the early years of the nineteenth century. Before gas became widely available at the turn of a tap, lighting was of low intensity and minimal coverage, limited to candles and oil lamps. The ability to light factory interiors long after dark not only improved productivity, but it changed working practices forever. But lighting turned out to be just one of the uses to which coal gas was successfully applied.

The initial application of coal gas to lighting is usually attributed to the Scottish engineer William Murdoch, who was also the inventor of the oscillating-cylinder steam engine amongst other things. But gas had actually been used six years earlier to light parts of the Fife mansion of coal magnate Archibald Cochrane, ninth Earl of Dundonald, so some credit for recognising the flammable qualities of coal gas – if not

Opposite: The retort house at Fakenham Gasworks in Norfolk dates from the early twentieth century, and contains one bank of eight retorts, the other of six, each with its own 'producer' or furnace. The workforce of eight men produced enough gas for around 500 properties. By comparison, the huge Beckton Gasworks in London employed 4,500.

Below: Boston Gasworks in Lincolnshire was rebuilt in 1875, the new facility with its 144 retorts surrounding twenty-four ovens being featured in the 26 March 1875 issue of *The Engineer*.

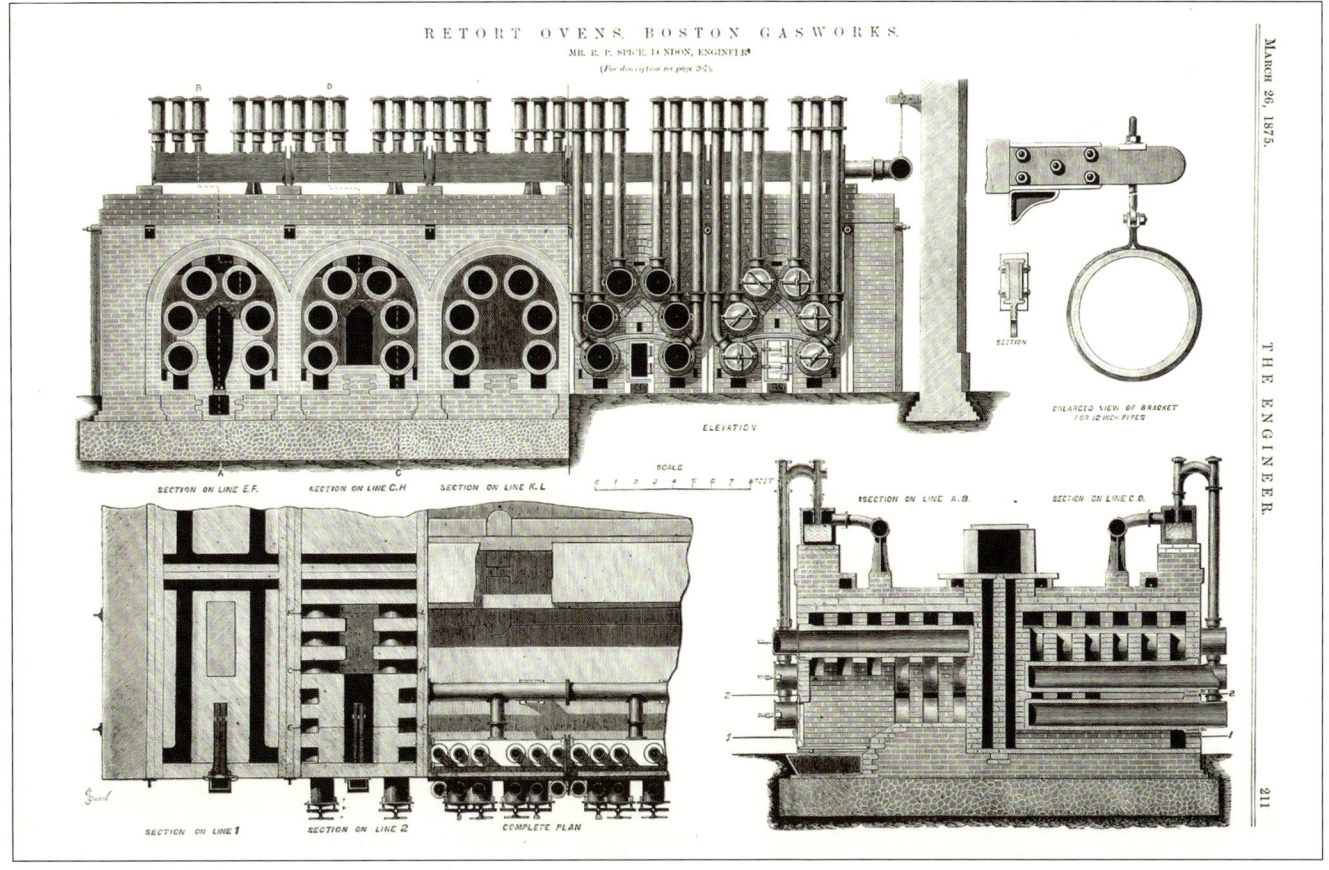

The 1912 30hp four-stroke Producer Gas Engine, built by the National Gas Engine Company of Ashton-under-Lyne and preserved at Hereford Waterworks, is prepared for action. It was originally installed in Alton Court Waterworks in Ross-on-Wye and was in constant use for fifty-two years, and used occasionally until 1970. It was designed to run on 'producer gas' – typically formed by passing air, or air and steam under pressure, through red-hot carbon, usually either coal or coke – but it currently operates on natural gas. The engine, designed for use where there were no gas mains, had its own little gasworks, and that is also displayed on site.

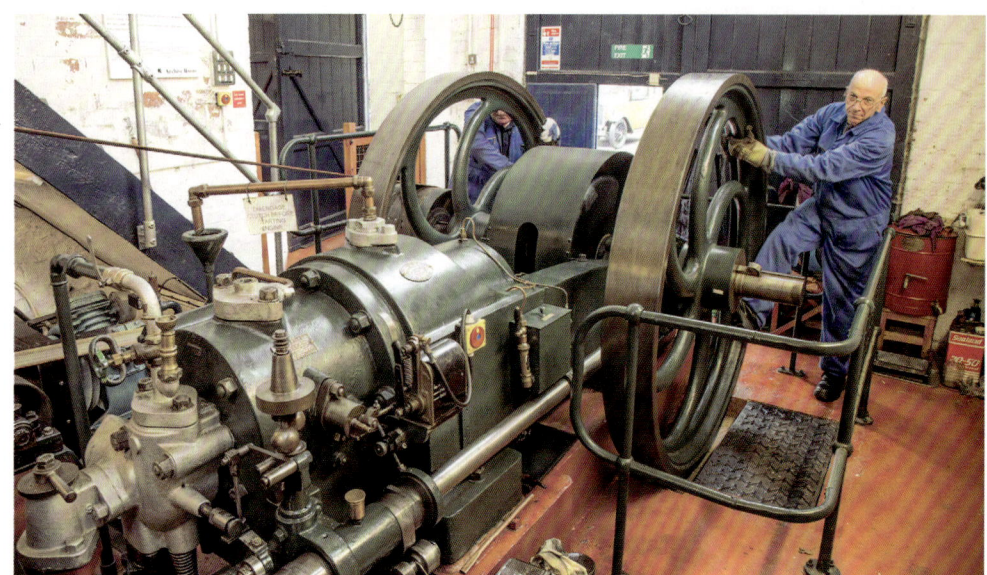

The heavy flywheels were essential for keeping a gas engine running smoothly.

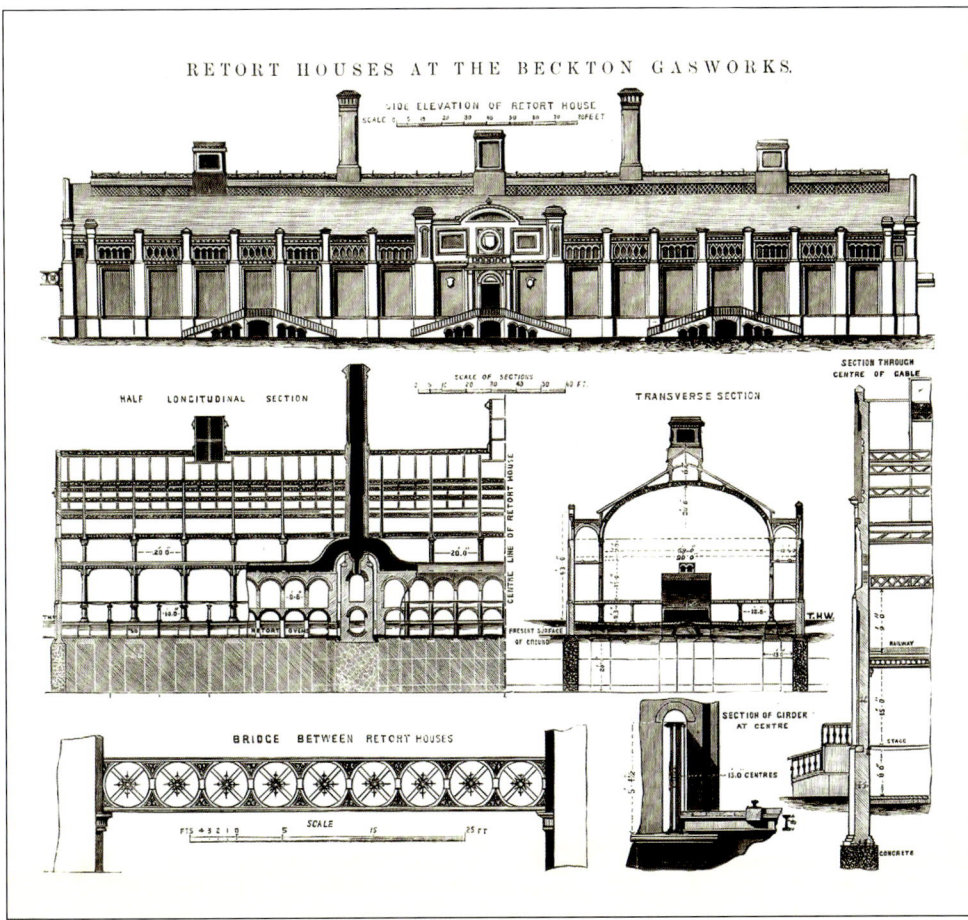

The Retort Houses at Beckton Gas Works, in an illustration from *The Engineer* in February 1870. At one time, this was the largest gas-producing plant in the world.

its potential – ought perhaps to go to him. However, igniting the gas and marvelling at its intensity was no more an entertaining sideshow to his process for distilling tar to make wood preservatives.

As with so many other areas of experimentation, parallel work was also being undertaken elsewhere by a number of inventors, so whether or not Murdoch was actually the first will remain open to debate. Certainly, by 1794 he had a working lighting system and a working methodology for the extraction of the gas from the coal itself.

Below left: So large was the Beckton Gas Works site that it had its own railway network. Locomotive Beckton No.1, a 0-4-0WT built by Neilson & Company in Glasgow in 1870, is preserved at Penrhyn Castle in North Wales.

Below right: A gas-fuelled engine at Fakenham Gas Works, used to pump gas around the site.

56 • THE GAS TRAMCAR

A PEEP AT THE GAS LIGHTS IN PALL-MALL.

Murdoch was one of the great innovators of the Industrial Revolution. He was the son of a Scottish miller, but had a passion for getting involved with steam. He made his way to Boulton & Watt's factory near Birmingham, and having been offered a job, worked for them for his entire career, introducing a whole range of improvements to their successful steam-driven beam pumping engines. He would actually spend a large proportion of his early years with Boulton & Watt living and working in Cornwall, to be close at hand whenever his expertise was needed with any of the engines.

To Murdoch certainly goes credit for the invention of the gas-holder – more usually referred to as the 'gasometer' – which was an essential part of any distribution system and would become a common sight in just about every town until the introduction of North Sea gas in the 1970s. Today only a few survive.

Murdoch's employment with Boulton & Watt was as an engine erector and he had an enviable reputation for being able to get maximum power out of their steam-powered beam engines for the minimum use of fuel. His experiments with gas, therefore, started out originally as little more than a sideline – a series of experiments conducted while he was in Cornwall, with gas from coal being just one of the flammable materials he was experimenting with – but by 1794 his Cornish residence was probably lit entirely by coal gas.

Given for whom he was working, it can be little surprise that the first commercial building to have gas lighting installed was Boulton & Watt's recently constructed Soho Foundry in Smethwick around 1800, followed by the installation of fifty gas lights into Philips & Lee's large Salford cotton mill in 1805 – the first large mill to be thus lit. That figure quickly rose to over 900, and as well as creating a cleaner source of light, gas lamps revolutionised and extended the working day.

On 28 January 1807, Pall Mall in London became the first street in the world to be lit by gas when street lights were turned on by the London and Westminster Gas Light and Coke Company. Later known as the Gas Light and Coke Company, and formally incorporated in London in 1812 – with an address, appropriately, in Pall Mall – it became a major driver in the development of gas and coke production, and its several gasworks around the capital produced millions of cubic feet per year. Beckton Gasworks, for example, provided a large proportion of all the household gas used north of the Thames in late Victorian London.

Opposite top: A cartoon by Thomas Rowlandson (1757–1827). The speech bubbles describe how gas lights worked and reactions to them.

Gentleman: 'The Coals being steam'd produces tar or paint for the outside of Houses – the Smoke passing thro' water is deprived of substance and burns as you see.'

Irishman: 'Arragh honey, if this man bring fire thro water we shall soon have the Thames and the Liffey burnt down – and all the pretty little herrings and whales burnt to cinders.'

Bumpkin: 'Wauns, what a main pretty light it be: we have nothing like it in our Country.'

Quaker: 'Aye, Friend, but it is all Vanity: what is this to the Inward Light?'

Street Girl: 'If this light is not put a stop to – we must give up our business. We may as well shut up shop.'

Client: 'True, my dear: not a dark corner to be got for love or money.'

Opposite bottom left: The surviving gas holder at Fakenham Gas Works in Norfolk. Gas holders, some many times this size, were once a common sight across the country.

Opposite bottom right: The light from the delicate gas mantle changed working conditions in mills and factories.

Left: A contemporary engraving of Boulton and Watt's Soho Works, the first to be lit by gaslight.

The Journal *The Engineer*, in its issue of 26 October 1860, described how the Lenoir engine worked: 'The machine itself is little more than an ordinary horizontal engine; but instead of being brought into operation by the expansive force of steam, motion is produced by the combustion of a mixture of gas and atmospheric air; and this combustion is brought about by the action of an electric battery, so that instead of steam, air is introduced alternately on each side of the piston. This air, if air it can be called – for it is atmospheric air mixed with ordinary gas – is heated and expanded after it is introduced into the cylinder, by the combustion of the gas which it contains, the gas being ignited by a spark from the battery. The poles of the battery penetrate the cylinder at each end, and the circuit alternates accordingly, as the mixture of air and gas is introduced to the one or the other side of the piston.'

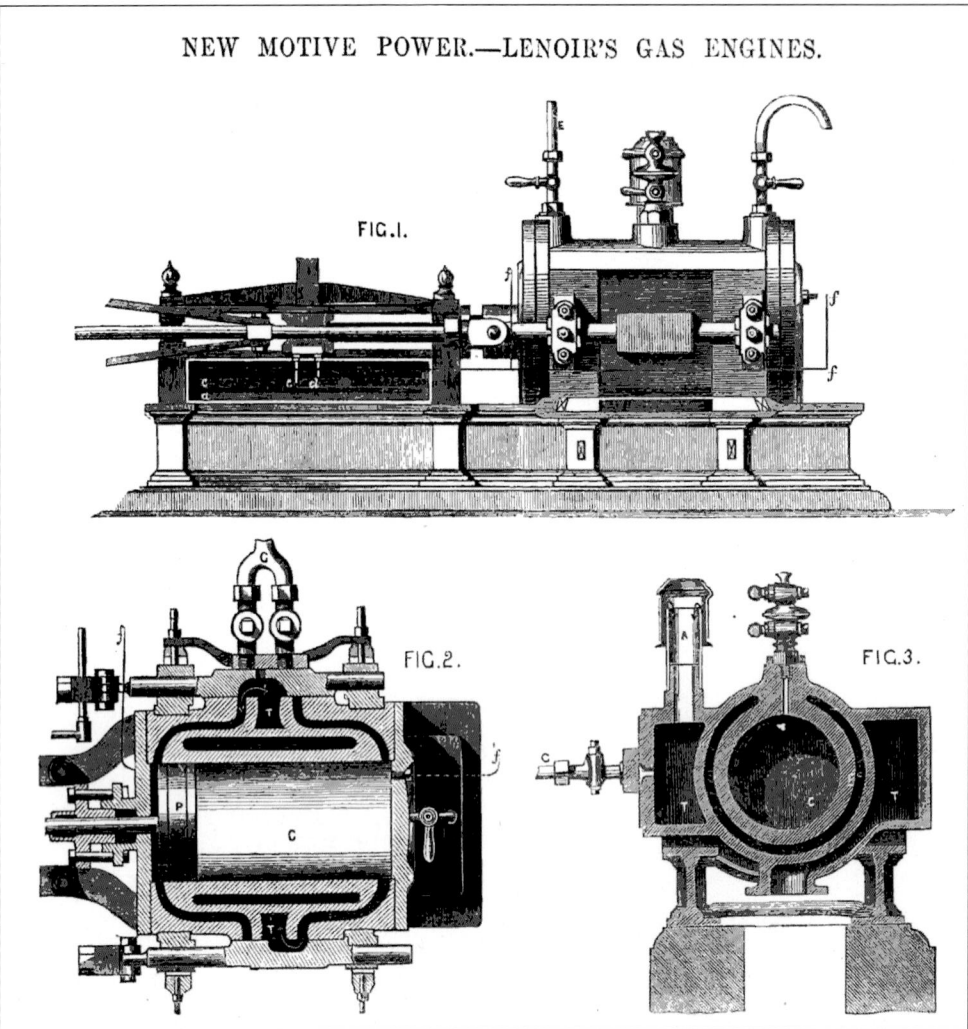

Below: The Lenoir engine as illustrated in the inventor's 1860 British and French patents from 24 January 1860 (France) and 8 February (Britain).

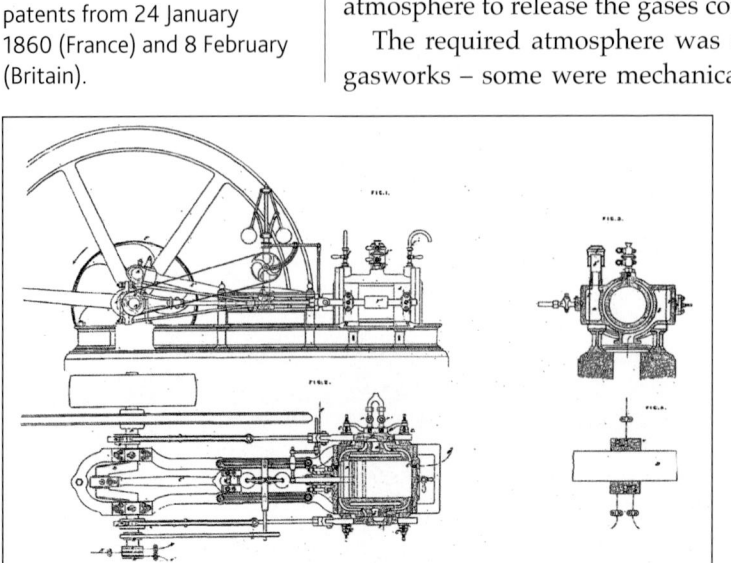

The production of coal gas was a simple enough process – albeit pretty unpleasant for the workforce – and basically involved the heating of coal in an oxygen-starved atmosphere to release the gases contained within that coal.

The required atmosphere was created in the retorts which were found in every gasworks – some were mechanically fed, but most small gasworks were hand-fed, the entire operation carried out round the clock by a surprisingly small workforce. In a small gasworks like Fakenham, for example, there were two furnaces – known as 'producers' – one heating eight refractory retorts, the other heating six. How many were in use at one time was dependent upon likely demand for gas, and the capacity of the gas-holder adjacent to the works.

The coal was shovelled into the retorts sequentially – or mechanically fed into the largest retorts in city gasworks – and the retort door sealed to create that oxygen-depleted environment. Over a period of hours, the gas was driven off and up the 'ascension pipe' and into a large 'hydraulic main' where it was given a primary wash to remove some of the tars. The water in the hydraulic main acted as a seal to prevent air mixing in with the gas when the retort doors were opened.

Through a series of washers and condensers, the gas was further cooled down to the temperature of the surrounding atmosphere, that process also helping remove tar and ammonia which were drawn off into a storage tank.

A further wash eliminated everything except hydrogen sulphide, which was removed by passing the gas through filter beds of iron oxide in tanks known as 'purifiers'. The total quantity of gas produced was then measured as it passed through the station gas-meter, and pumped into the gas-holder to await demand and distribution.

The most important by-product of coal-gas manufacture was coke, which had a multitude of uses. It replaced charcoal in iron smelting furnaces, and would later prove essential in the production of the millions of tons of steel which drove Britain's industrial expansion in the nineteenth century. But at small gasworks, much of the coke was drawn straight from the retorts and fed into the furnaces below, providing the fuel to keep the process cycle going.

From the outset, gas was seen as being a 'clean' fuel, whether it be used in a domestic environment for lighting, heating and cooking – certainly much more so than the coal fire and oil or candle lighting. While the exhaust smoke and fumes from coal burning were obvious to everyone, burning gas – with its invisible fumes and no soot and ash – seemed an enormous step forward.

In industry, it was plentiful and cheap, providing lighting of hitherto unimagined brightness and requiring many fewer people to operate gas-fired equipment than with coal. Its many by-products were, in some cases, even more valuable than the gas itself. Turning coal into coal gas while also marketing every chemical residue left by the process was a much more efficient use of it than simply burning it in a furnace to generate steam.

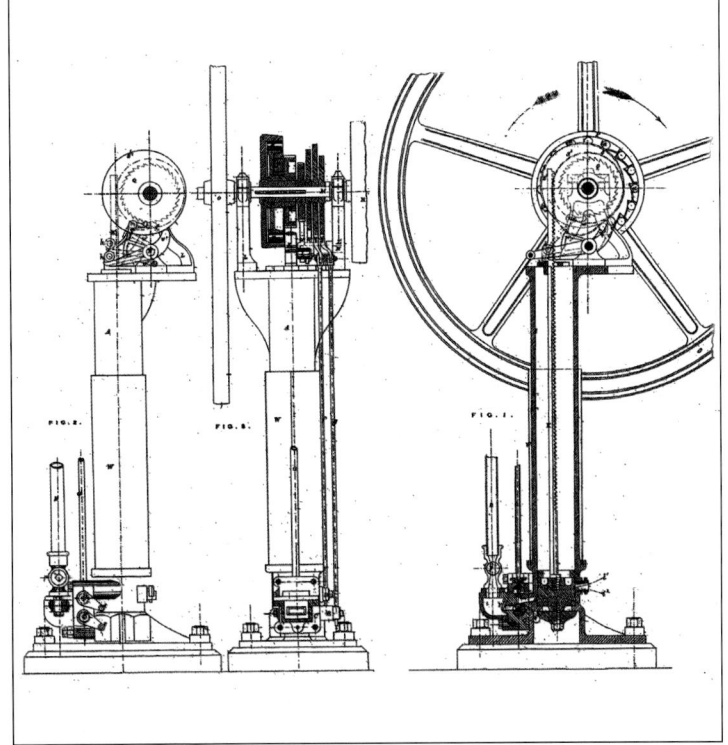

One of the patent illustrations from Langen and Otto's single cylinder vertical Atmospheric Engine, 1866.

It was 1860 before the first gas-fuelled engine was developed. There had been many experiments using hydrogen, coal gas and other gases to fuel engines before then, but credit for developing the first practical internal combustion engine to burn coal gas goes to the Belgian engineer Jean Joseph Étienne Lenoir in 1860. His engine was not sufficiently powerful for most practical applications and used a lot of fuel, but despite its limitations, the journal *The Engineer* recognised its potential and understood that such engines marked a major advance in modern power systems.

In a world where steam power and water power were king, the gas engine was sufficiently unusual as to warrant extensive reporting. As with so many inventions, describing them before specific terminology had evolved was often clumsy:

> 'There are many, and even domestic, manufactures in which it is not possible to make use of the steam engine, and where, in consequence, it is necessary

to have recourse to manual power – the most expensive of all powers – where this new machine, from its simplicity of action, and the readiness with which it may be put into operation, will be found of considerable service. Without boiler, without fuel (in the ordinary meaning of the word), without a chimney to promote a draught and to conduct away the products of combustion, this machine, no doubt, starts with many advantages.'

With improvements, it did, however, enjoy some success running small machinery and, more importantly, it alerted others to the potential of gas as a fuel and it would be one of those later inventors who developed a truly practical gas engine.

Just seventeen years after Lenoir's first success, Nicolaus Otto, whose achievements would prove central to the history of the gas tramcar, developed and patented his first 'atmospheric' gas engine in February 1866, using a naked flame to ignite the mixture of gas and air in the engine's combustion chamber.

In 1876, Otto built his first compact four-stroke gas-fuelled engine. He conceived it as a stationary engine, but when smaller versions of his design were developed, able to run on a variety of gases including 'town gas', others – amongst them his general manager Gottlieb Daimler – saw the engine's potential in powering vehicles. That difference of opinion would see Daimler and Wilhelm Maybach leave the company and set up on their own – becoming major players in the development of automobiles and the petrol-fuelled internal combustion engine.

Above: Nicolaus Otto was featured on a West German postage stamp in 1952 to mark the seventy-fifth anniversary of the introduction of the original Otto-cycle gas engine, and the beginning of the dominance of the internal combustion engine.

Opposite: The Otto/Langen gas engine at the Gasmotoren-Fabrik Deutz headquarters in Cologne carries serial No.1. It is a vertical free piston atmospheric engine, made from cast iron, steel and bronze. Another example is preserved in Sydney.

Left: The pair of 75bhp gas engines built by Crossley Brothers of Manchester which generated the electricity for the motors which drove the Widnes–Runcorn Transporter Bridge's gondola across the Mersey. They also charged 245 wet cell chloride batteries to provide back-up power. The bridge and its engine house were described and illustrated in the 16 June 1905 issue of *Engineering*. The gas supply was piped in from Widnes gasworks.

Right: Two illustrations of his single-cylinder four-stroke engine from Nicolaus Otto's patent application *Improvements in Gas-Motor Engines*, filed in January 1876. This was just one of a progressive series of patents filed between 1876 and 1879.

Below: A single-cylinder four-stroke Otto-cycle engine, by Schleicher, Schumm & Co of Philadelphia, built two years after Otto patented his engine. The Philadelphia company built its first slide-valve gas engine in 1878 as the American licensee of Gasmotoren-Fabrik Deutz. Their factory was known as the Otto Gas Engine Works.

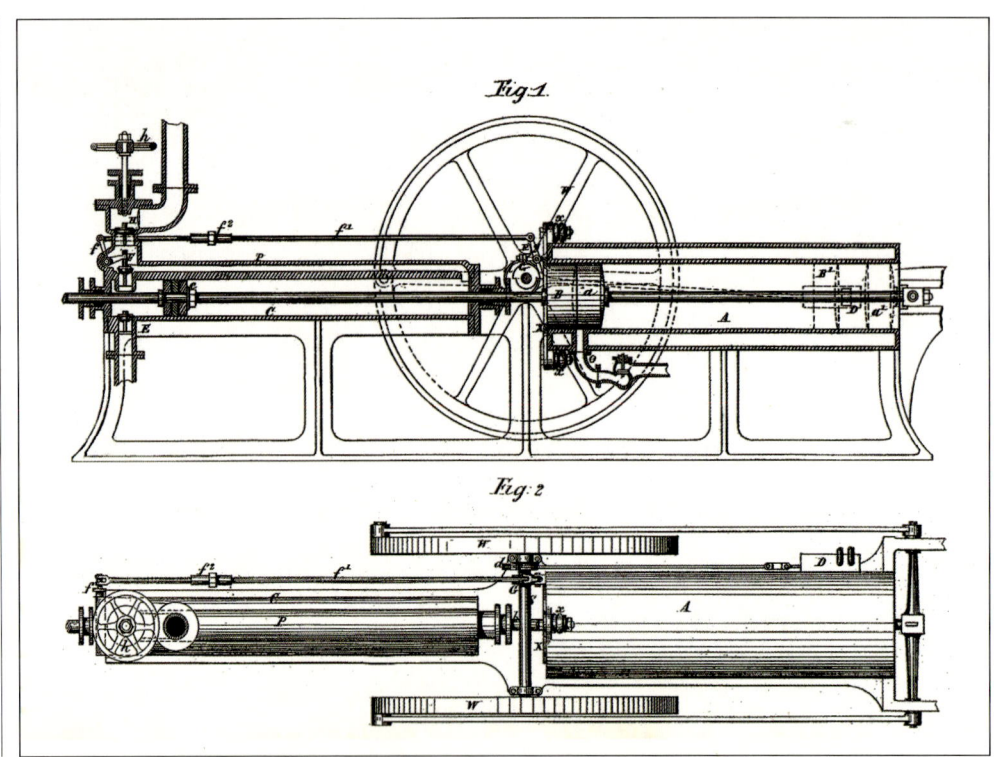

THE GAS REVOLUTION • 63

Left: In this 1895 Hot Tube Gas Engine by the Campbell Gas Engine Company of Halifax, now preserved at Hereford Waterworks, town gas was ignited as it was fed under compression into the combustion chamber.

Below: This 150hp gas engine, built in 1915 by Crossley Brothers and now preserved in Sheffield's Kelham Island Museum, is said to be the largest single cylinder gas engine ever manufactured by the Manchester company. It drove a rod and bar rolling mill in Sheffield. Established in 1867, Crossley engines were used in many factories across the world.

Right: A postcard of the Gasmotoren-Fabrik Deutz factory c.1920, by then one of the largest engine plants in the world. Cologne Cathedral can be seen in the distance.

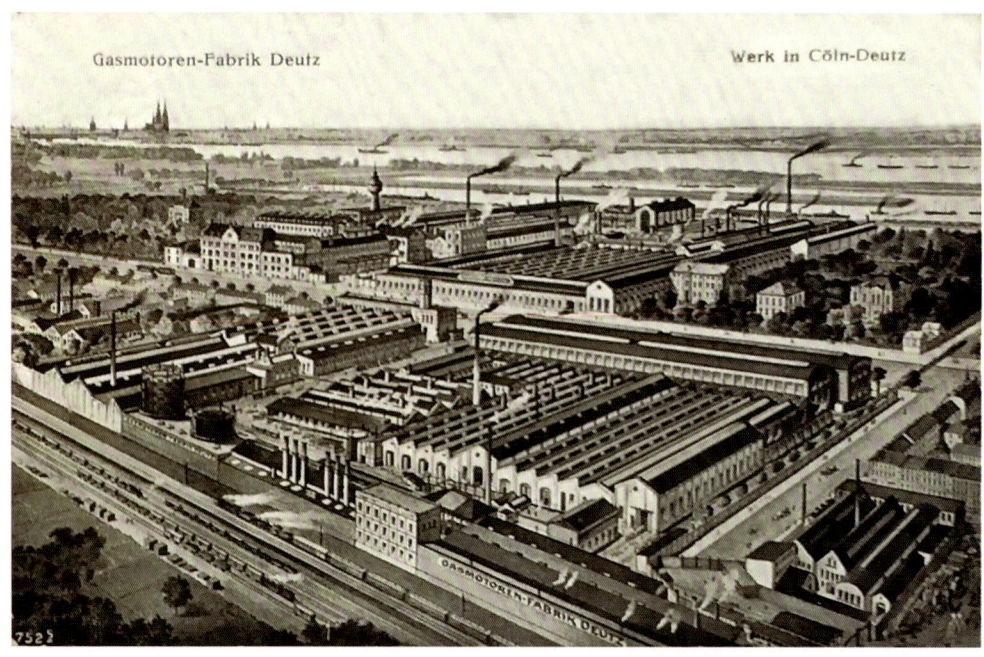

Below: A large industrial installation of gas engines built and installed by Tangye Brothers of Smethwick, one of the world's biggest builders of steam, oil and gas engines. Tangye's large and efficient gas engines were used to drive many large electricity generating sets.

By the time the first 'Otto-cycle' gas engine was produced, Otto's company was based in Deutz, a suburb of Cologne, and trading as Gasmotoren-Fabrik Deutz. The design was interesting and a little different from the internal combustion engine as it would evolve in the decades which followed. Rather than a simple explosive ignition, the Otto engine delivered the gas under compression and relied on a longer slower burn of the fuel to create the power.

THE GAS REVOLUTION • 65

Left: A Tangye Hot Tube Gas Engine at the Black Country Living Museum. These engines could be run using 'town gas' or vaporised oil. Ignition was by passing the gas/air mixture through a red-hot metal tube where it exploded under compression. There was no spark ignition system on these engines so they needed to be 'bump-started' by hand, but they proved to be efficient and easy to maintain.

The first engines delivered about 0.5 horsepower, but by the end of the nineteenth century, very large 1,000hp Otto gas engines were being built.

Otto never lived to enjoy the real success of his engine, sadly, having died in 1891 at the age of only fifty-eight, long before its full potential was realised.

Early in the evolution of the gas engine, its potential application to vehicle traction was understood by several inventors. All that was needed was for the power output

Below: This patent, 1883-built six-stroke engine by Samuel Griffin of Bath used his modification of Otto's four-stroke cycle. Built by Dick Kerr & Company, it operated for seventy-three years at Beckton Gasworks until withdrawn in 1956. It is now displayed in The Museum of Bath at Work. The only other example can be seen working in the Anson Engine Museum in Cheshire.

to reach a practicable level, and reaching that potential took a lot longer than many had anticipated – especially considering that the potential of gas engines to power vehicles had been recognised in the closing years of the eighteenth century.

Writing in his 1885 book *Les Moteurs à Gaz*, Gustave Richard enumerated some of gas's advantages over other fuels

> 'Le moteur à gaz présente comme locomoteur quelques avantage particuliers. Plus leger, à puissance égale, que l'air comprimé, puisqu'il permet d'ajouter au travel de sa compression l'energie considerable de sa combustion, le gaz constitue, sous cette forme, l'agent de locomotion de la force la plus énergique dont nous puissions disposer, puisqu'il fournit, à poids égal, plus de deux fois plus le travail que le charbon.'

In other words, power for power, the gas engine was lighter than either compressed air or steam, and was, he said, capable of twice as much work as a similarly sized steam engine. He believed it to be ideally suited to vehicle traction.

Of particular interest was the application of gas engines to tramcars proposed by the Prussian engineer Conrad Krauss in 1881 which were driven by a single-cylinder horizontal gas engine using a mixture of gas and air, both stored on the vehicle at ten atmospheres. The patent application was made on behalf of Krauss, described as an engineer 'of Hannover in the Kingdom of Prussia' for *A Gas-locomotive for Tramways and Railways of Secondary order*, by his Agent, Carl Pierper, and became British Patent No. 309 in January 1879.

Richard also commented on the early proposals of Holt and Crossley in 1879 (*see* the chapter on gas trams in Britain), the London-based civil engineer father and son Joseph and Joseph Quick – much better known for their work designing and building waterworks – whose engine was patented as British patent No.5575 in December 1881.

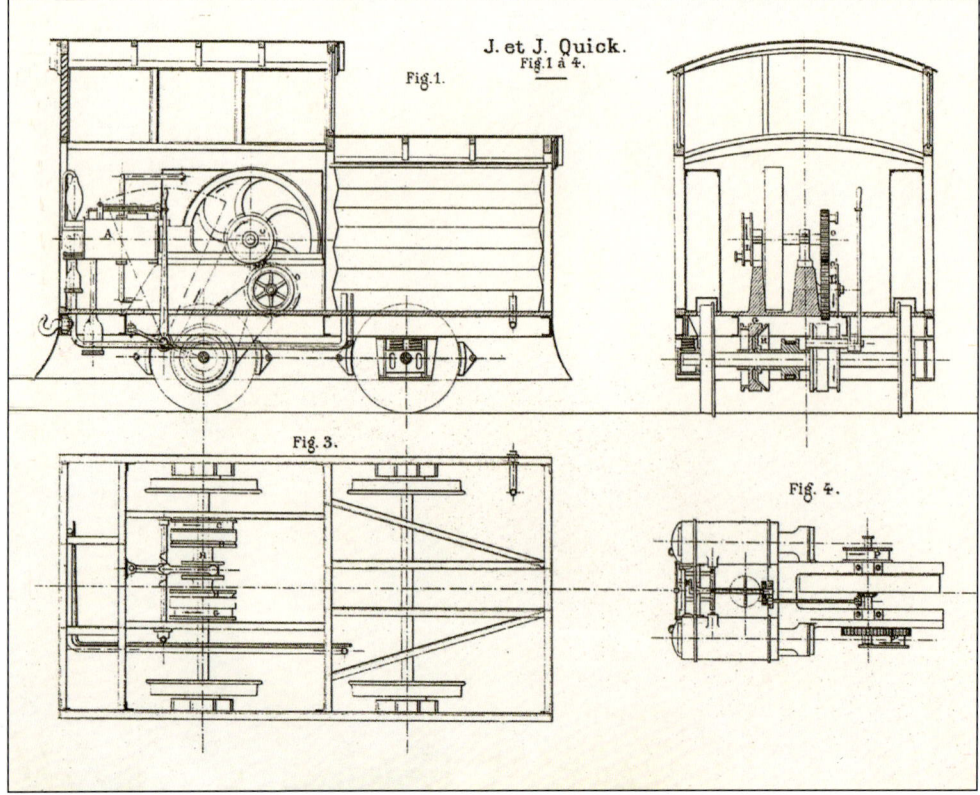

Father and son Joseph and Joseph Quick, better known as water engineers, patented their design for a gas-engined tramway locomotive – intended to haul former horse trams – in 1879. It seems never to have been built.

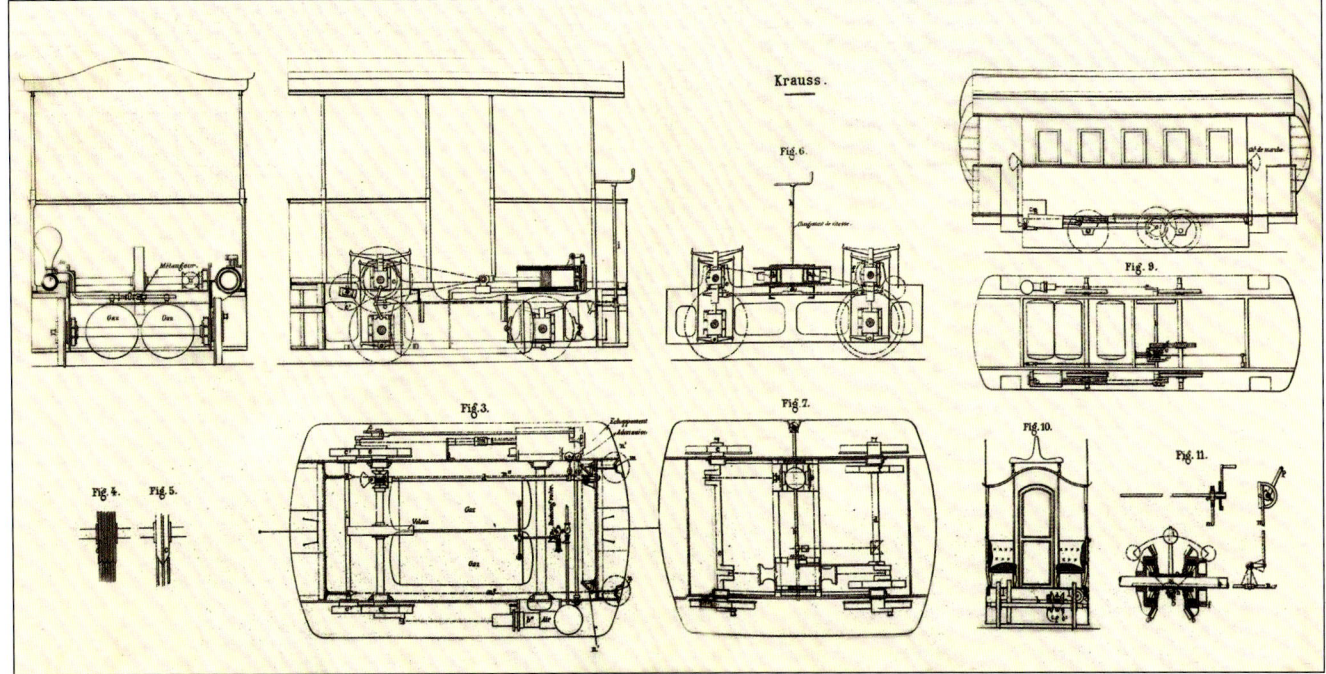

Typical of the laxity of patent law at the time, they all patented aspects of the gas tramcar which would later also be patented by Carl Lührig, the man who actually made gas trams a reality – horizontal engines, friction clutches, reversing gears and so on – but Lührig (*see next chapter*) would not even file his first patent until 1891.

As far as it has been possible to ascertain during the course of this research, none of these vehicles was ever actually run on tramways. Had they been, in every case the size of the engines proposed would have proved ineffective on anything other than a completely level track.

Richard also listed the French engineer Marcel Deprez, better known for his pioneering work in electricity transmission, and Marchant and Wrigley who in 1883 proposed their own gas tramway system.

Just two years after Otto was granted patent protection for his four-stroke gas engine, the first provisional patent application for using such an engine to power a tramcar was submitted on 11 January 1878. However, that engine was never built.

The engine which would be developed for use in trams was a compact horizontally opposed two-cylinder design initially producing around 7hp and using Otto's four-stroke cycle. The four-stroke cycle involved: intake, where gas and air enter the combustion chamber; compression of the mixture by the piston; ignition and combustion of the gas/air mixture; and finally exhaust. The engine, even in its earliest and most basic form, was quieter than Lenoir's, burned more efficiently and, from a commercial point of view, used a lot less fuel.

The British licence to manufacture Otto-cycle engines was acquired by Crossley Brothers of Manchester and became one of their most important product ranges.

As manufacturing tolerances improved, the power of the gas engine was gradually increased and its application to motive power became a reality. It did, however, take some time before the 12hp and 15hp horizontally opposed twin-cylinder engines which would be used in a number of gas-fuelled tramcars were developed.

While the Quicks only proposed a gas-engined locomotive, Conrad Krauss of Linden, Hesse, also planned an integrated vehicle with the single cylinder horizontal gas engine beneath the floor of the passenger cabin. Both options were illustrated in Gustave Richard's *Les Moteurs à Gaz* published by Dunod in Paris in 1885. Monsieur Richard described himself as 'Ingénieur Civil des Mines'. Krauss's designs were patented in 1879 as British Patent No.309 *A Gas-Locomotive for Tramways and for Railways of Secondary Order*. The provisional specification was filed on 8 October 1878 and the patent granted the following year. As with the Quicks, there is no evidence that any of these vehicles were ever constructed.

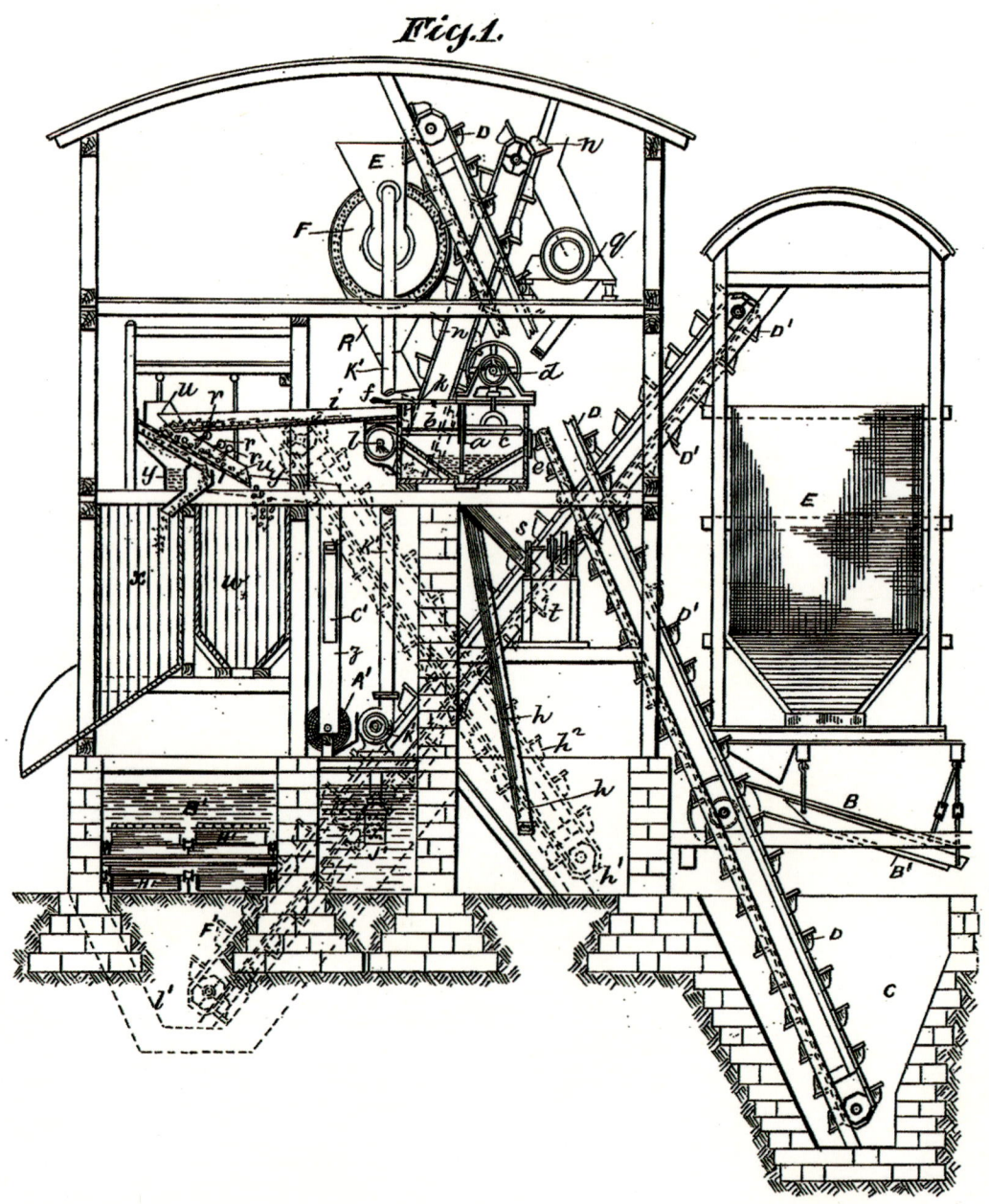

THE FIRST GAS TRAMCARS

As several inventors had recognised, one of the key steps along the way to turning a stationary gas engine into a power unit capable of moving a vehicle was the development and combination of a friction clutch and gearbox. The Dresden-based engineer Carl Lührig (1840–93) was certainly not the first to patent such a system, but he was the first to combine all the necessary elements in an operational street tramcar system that would carry fare-paying passengers.

One of Lührig's companies – Lührig Coal and Ore Dressing Appliances Limited – was based in Britain, with headquarters in Westminster, and was well known long before the idea of placing a gas engine into a tramcar was ever considered. The company's reputation was built not on transportation systems, but on key aspects of the coal and mineral ore processing industries. Indeed, he described himself primarily as a 'Mining Engineer' and developed some very heavy machinery.

Lührig System coal and ore washeries and jigs were installed in mines and collieries around the world – with many in Australia – and the company held many patents for post-mining equipment, including two in 1892 – registered in both Britain and the United States – *Apparatus for washing, separating and concentrating ores of different specific gravity* and *Separating and cleaning coal and other minerals*.

The Lührig 'Piston Jig' was used to separate materials of different density and specific gravity – in coal washeries, for example, heavier elements sank to the bottom of the water-fed jig while the less dense lumps of coal, buoyed by the fast-flowing water, were cleaned and then skimmed off.

Opposite: Carl Lührig had already established an enviable reputation in the world of mining machinery before he became involved with the development of the gas-engined tramcar.

Left: Lührig's Dessau tramcar No.7 with its crew, an illustration from the American *Cassier's Magazine* in June 1895. Car No.7 was one of the initial batch of nine small cars capable of carrying twenty-eight passengers, fifteen of them seated.

70 • THE GAS TRAMCAR

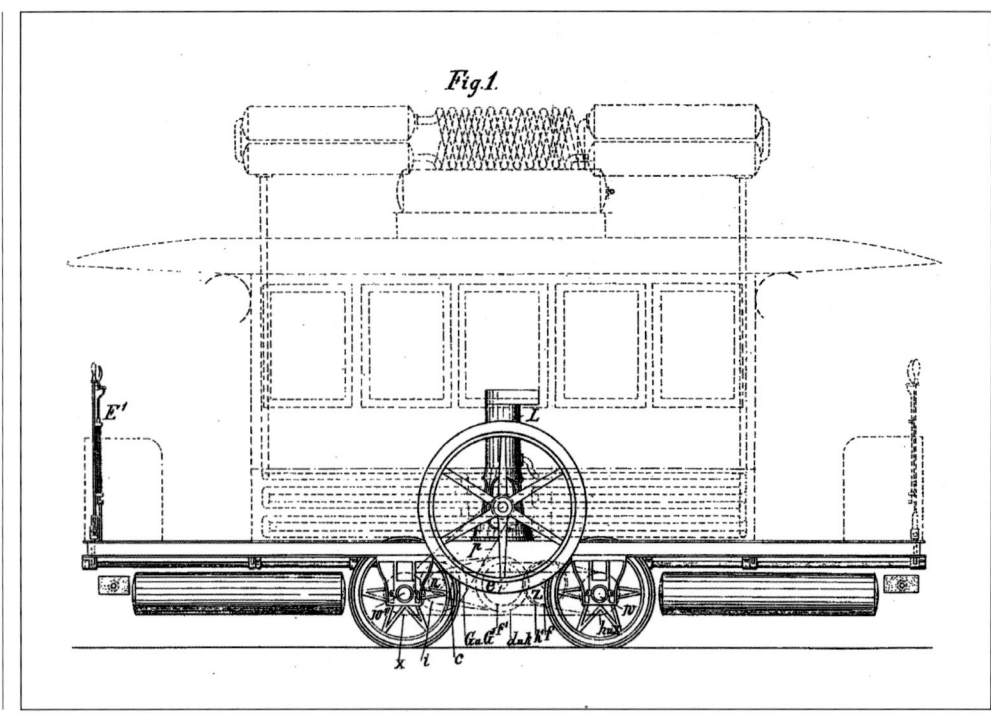

Right and below: Lührig's original plans for his tramcar's engine layout as envisaged in his 1891 Patent No.15,655. He proposed using two single-cylinder gas or petroleum engines mounted vertically. The fact that this patent suggests a 'Blessing engine' might be suitable perhaps implies that Lührig was not aware of the compact four-stroke Otto engine at that time.

Below right: Oskar Blessing's 1887 vertical single-cylinder engine – Saxony Patent No. 44152 *Tram Car* – had rotative motion delivered to the tram's wheels by a geared chain drive.

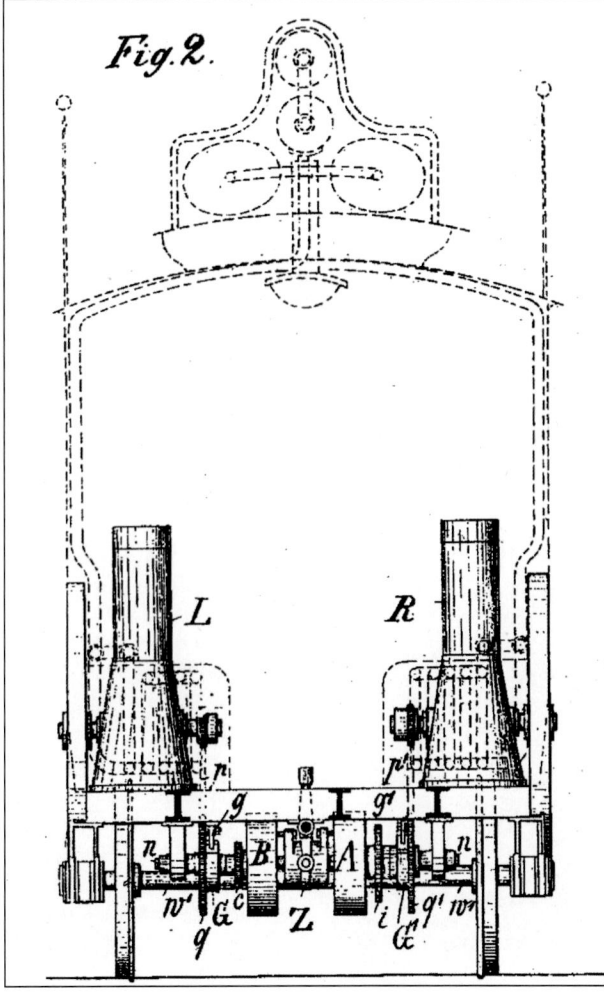

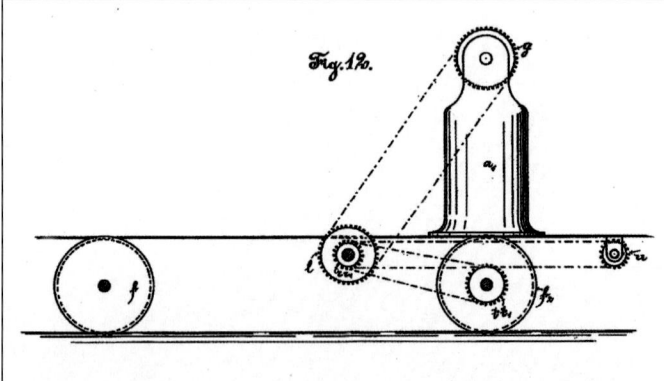

Other Lührig devices were designed to grade the coal according to size, but what exactly made him turn his attention to tramcars is unknown.

Carl Lührig designed his first gas trams in the closing years of the 1880s, later developing systems to operate in Dresden and elsewhere. His first cars differed considerably from those which were later introduced in Dessau and in Britain. The evolution of the tram design was chronicled and illustrated in a series of eight British Patents, the first submitted to the Patent Office in September 1891, No.15,655, the last in June 1893, No.11,506.

In a relatively short period of time, he must have spent a great deal of money on patent applications, as each time he revised his thinking about how best to develop a commercially viable vehicle, he sought to protect those developments.

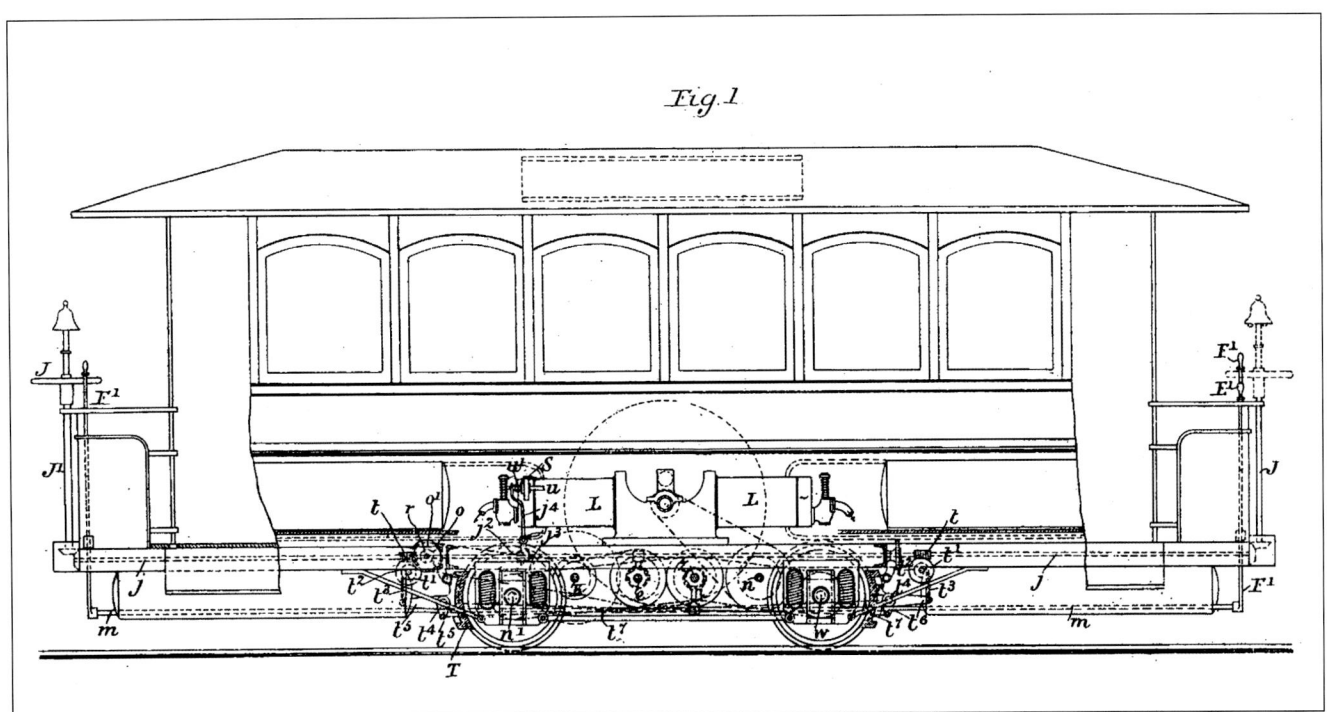

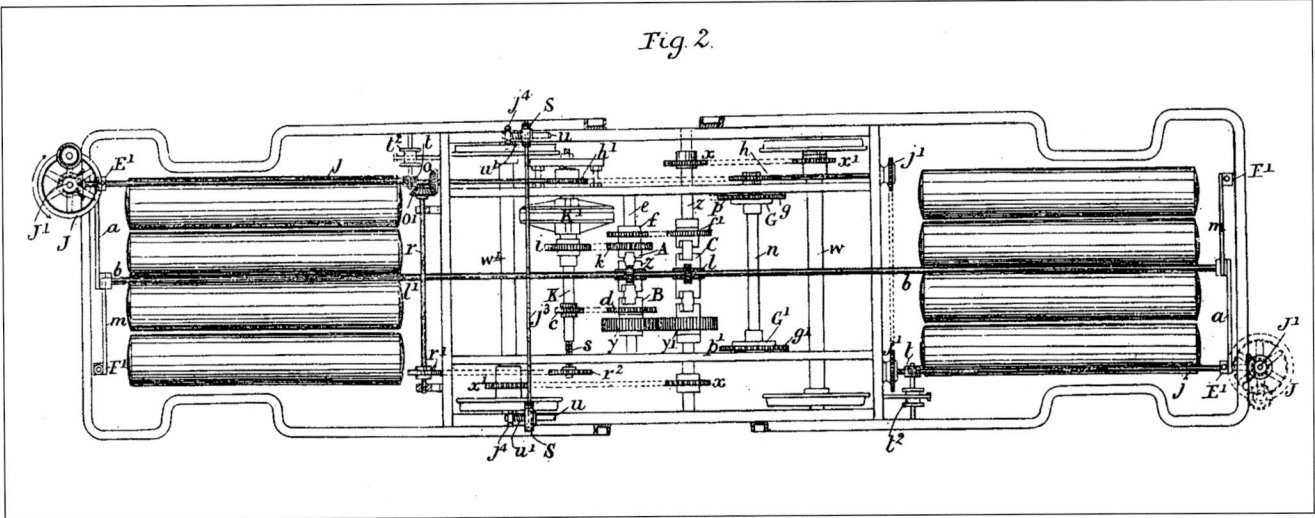

Top and above: By the time of Lührig's second patent, December 1891, a larger vehicle with two-cylinder horizontal engines under the seats was envisaged, equipped with frictional clutches to improve braking and stopping.

In all but one of his patents, and starting with *An Improved Construction of Locomotive Vehicles*, submitted on his behalf by Abel & Imray, Patent Agents of Southampton Buildings in Chancery Lane, London, Lührig described himself as a Mining Engineer and gave his address as Reichstrasse No.28, Dresden. The engine he chose to power the vehicle was very different from that which would drive his first successful trams.

Power came from two vertical single-cylinder gas engines, one either side of the vehicle – severely restricting the number of passengers who could be accommodated – and the fuel was carried in pressurised gas bags beneath the floor either side of the tram's very short wheelbase. Cooling water was carried in tanks on the roof, and the engines were fitted with simple silencers to muffle their noise.

While he did not offer an engine specification in that patent, he did suggest a possible option, albeit one for which only relatively few records have been located in the course of this research:

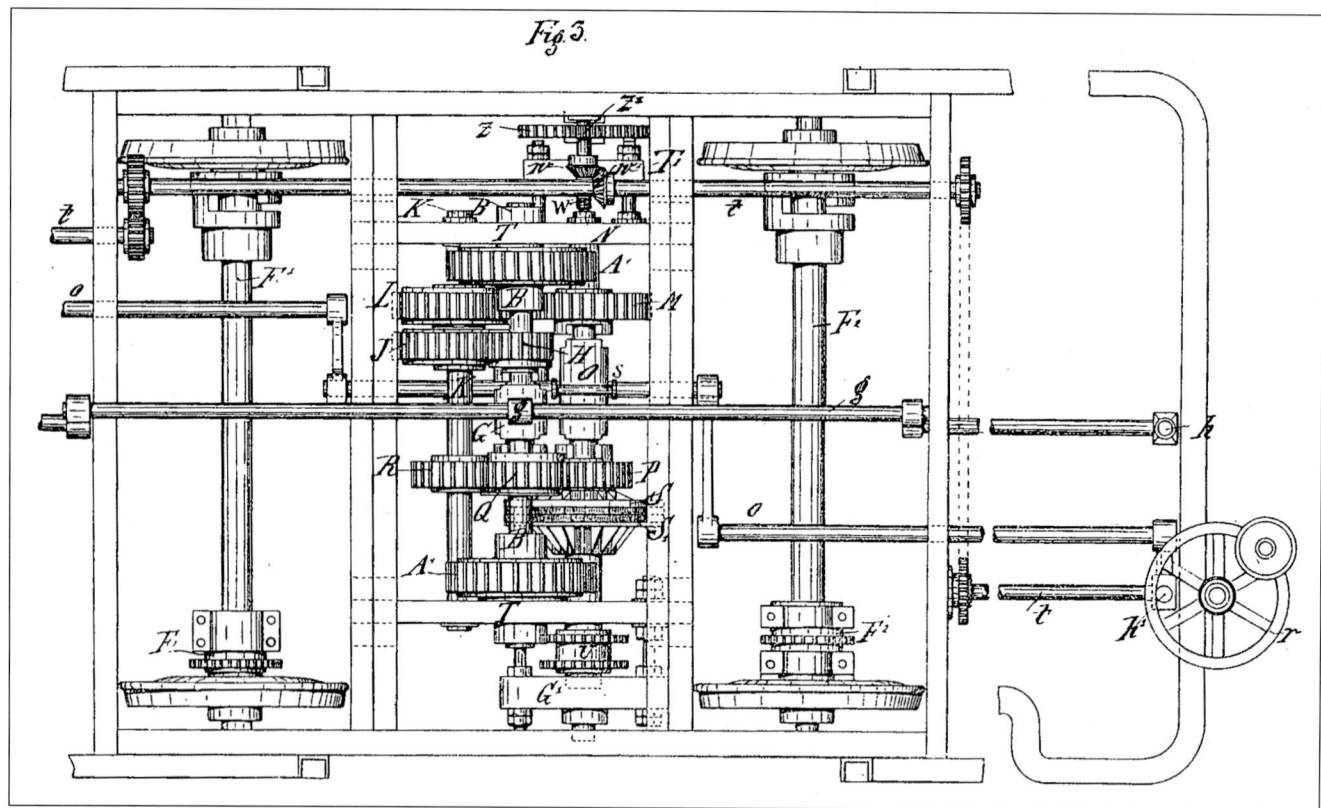

The tramcar's complex transmission and gearing as illustrated in Lührig's March 1892 patent application, No.5707. The motors were to sit on frames above the gearing, held in place by four bolts. To carry out repairs those bolts simply had to be undone allowing the entire frame to be withdrawn. The intention was that a replacement engine on its frame could be fitted just as easily.

'Of the gas and petroleum motor engines of known construction, that of Blessing is particularly suited for the purpose, because it offers in a very compact form, a source of power that can be worked at an exceedingly cheap rate.'

Contemporary with Lührig, the German engineer and inventor Oskar Blessing had patented a gas engine in his native Saxony in December 1887 – Patent No.44,152 – and fitted it to a former horse-drawn tram which he demonstrated in the Anger-Crottendorf district of Leipzig. While it clearly impressed Lührig, Blessing's tram was not developed further, despite also being patented in America (US Patent 391,774, 30 October 1888).

The engine was a vertically mounted single cylinder engine connected to the front axle of the tramcar by a chain drive, and it did not have the distinctive large flywheel which characterised Otto's successful engines.

By the time he submitted his second patent application in December 1891 – Patent No.21,121 also titled *An Improved Construction of Locomotive Vehicles* – Lührig envisaged an altogether larger tram car with an extended wheelbase, and the two vertically mounted engines had been replaced by a pair of horizontal engines of unspecified design.

It is apparent from the wording of the third patent application, No.5,707 *Improvements in Tramway Vehicles Driven by Motor Engines* dated 23 March 1892, that Lührig already had some of his cars in service, and was receiving feedback from passengers – not all of it entirely favourable, for he noted in the Provisional Specification that:

'It is a well known disadvantage that with tramway vehicles that are propelled by gas or petroleum motor engines the unavoidable smell from the combustion products easily passes into the carriage and becomes a nuisance to the passengers.

It has therefore always been necessary in such carriages either to provide a separate compartment as engine room, or even to employ an entirely separate vehicle as locomotive. The cumbrous appearance and the great expense of construction and working caused by such arrangements, more particularly on account of the necessary turntables or reversing sidings, have materially hindered the extended use of such tramway vehicles driven by motor engines.

According to the present invention, the motor engines and driving gear have been so arranged that nothing is to be seen from the outside, the motor engines (two twin engines) being housed beneath the two large longitudinal seats of the vehicle, while their fly wheels are contained within the hollow sides thereof. By such arrangement it has been possible to entirely cut off all communication between the engines and the interior of the vehicle, in such manner that neither smell nor noise can penetrate into the latter.'

He may have reduced the levels of noise and smell, but he had clearly not eliminated them as passengers would continually report that both issues were considered to be disadvantages to using the trams. In a nod to the unreliability of early engines, that third patent also envisaged the power units being easily removed by sliding them out through the side maintenance doors and inserting replacements, thus enabling repairs to be carried out in the company's workshops without removing the vehicle from service.

Above: Lührig's fourth British patent, No.11,897, 25 June 1892, was titled *Improvements in or relating to the Wheels and Axles of Tram-Cars*. The design described and illustrated in the patent enabled 'the independent acceleration or retardation of motion and for the simultaneous longitudinal displacement of the wheels' as the tram went round tight curves, using a spring loaded 'sleeve or clutch' on the axle.

Left: Lührig's Dessau tramcar No.6 with its inspection and maintenance doors open to show the flywheel, as illustrated in *Cassier's Magazine* in June 1895.

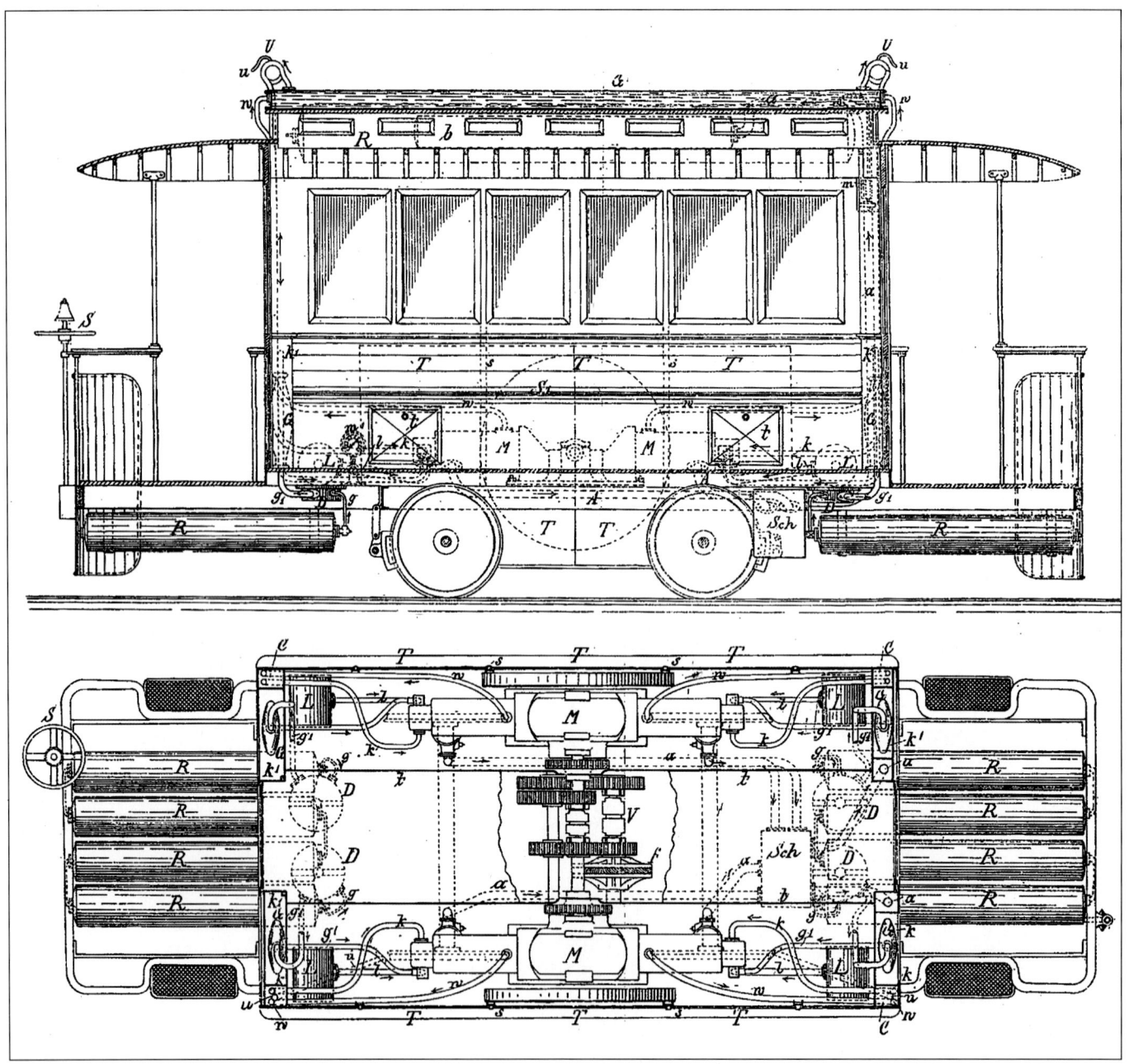

Two views of the significantly larger design of tramcar proposed in Lührig's fifth British Patent – No.15,841 – submitted on 3 September 1892 and accepted just a month later.

For his next Patent, No.11,897, *Improvements in or relating to the Wheels and Axles of Tramcars*, submitted in June 1892, a different Patent Agent was used, this one being Alfred Julius Boult of 323 High Holborn.

The patent itself comprised the introduction of clutches on the wheels, giving those wheels a degree of independent movement – despite being mounted on fixed axles – in order to allow the tram to negotiate tight curves more easily, avoiding what Lührig described as the 'fast wheels' being 'dragged along' as the vehicle negotiated curves.

By the time Abel & Imray submitted Lührig's next application – which would become Patent No.15,841 – the layout of the vehicle had evolved considerably and now had a pair of two-cylinder horizontally opposed engines – a design of engine which would become standard in later trams, albeit most using just one rather than two.

The engines were located beneath the seats, their gas supply contained in eight 'caouttchouc bags' – unvulcanised natural rubber – in two groups of four in metal containers beneath each end platform. Two more bags were mounted on the roof.

On the roof there was also a tank of water to cool the engines. The use of rubber bags which could only be pressurised to around eight atmospheres limited the amount of gas which could be carried.

In common with horse-drawn and steam-hauled trams, Lührig's earliest designs only envisaged there being a driving station at one end of the vehicle, which would have required turntables or turning triangles at either end of each tramway. Production designs rectified that omission.

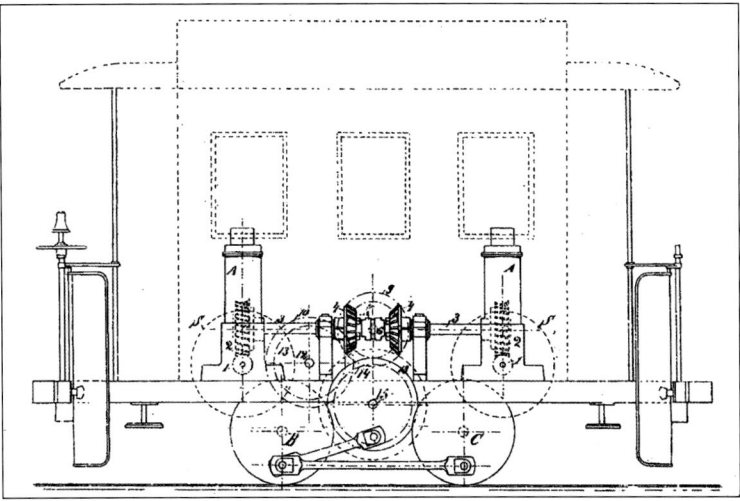

Lührig's drawing for a separate twin vertical-engined locomotive unit, from Patent No.19,070.

The early cars were quite small and heavy, having two Otto-cycle or similar engines – each with a large heavy flywheel. The two engines working together reportedly caused the car to vibrate excessively, giving the fourteen seated passengers and six standing on each end platform an uncomfortable ride.

In any event, such a small carrying capacity was not commercially sustainable and his first attempt to improve profitability was to dispense with one of the engines, and it was such a vehicle, or one very similar to it, which was used on the Dessau tramways, and which was first tested on British rails in Croydon in 1893. These vehicles were notoriously under-powered and Lührig next considered either fitting larger and more

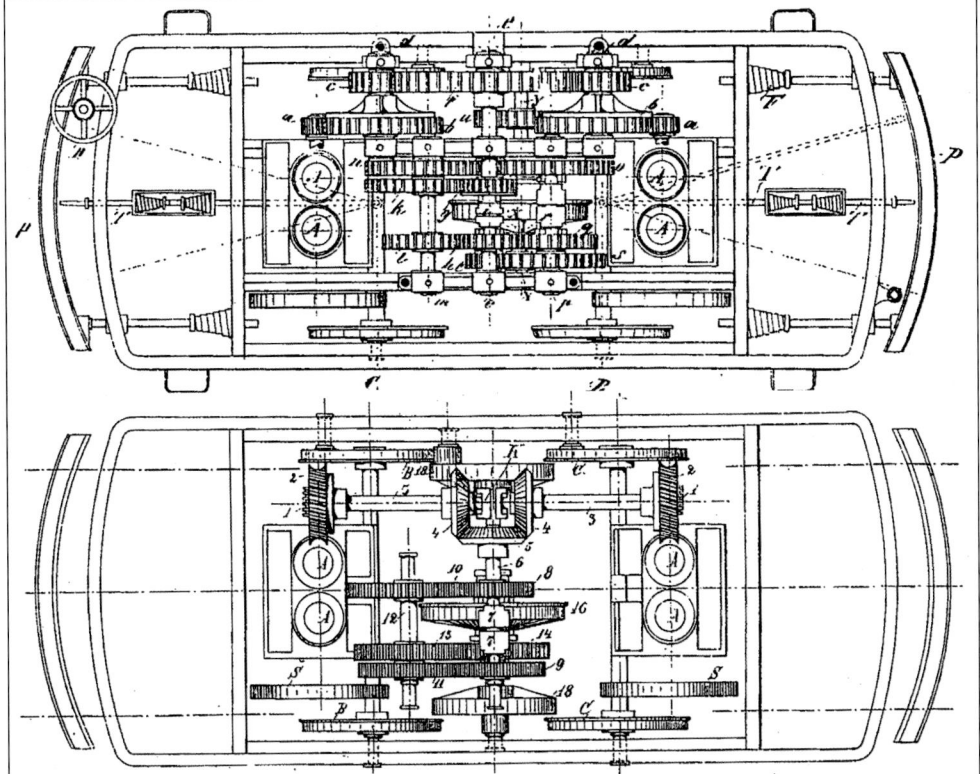

Two illustrations of Lührig's proposed twin vertical-engined locomotive unit, patented in 1892. A direct replacement for horses, at least three variants of this vehicle were built between 1892 and 1898, two of which can be seen on the following pages.

powerful engines into bigger vehicles – should such engines be available – or follow the same route which steam and compressed air systems had adopted – separate power cars.

That was the essence of Patent No.19,070 *Improvements in Locomotive Engines or Vehicles for Street Tramways*, filed on 24 October 1892 and granted nine months later. Here, Lührig reverted to a pair of transverse-mounted vertical twin-cylinder engines and some elaborate gearing.

Much was made in the patent of the fact that the design of the transmission system could equally be applied to engines fuelled by 'either gas, petroleum, compressed air or other fluid pressure' – protecting his invention against future developments in engine design and, according to the specification:

> 'There has not, up to the present time, been produced any propelling engine for tramways in which the machinery has been rendered entirely invisible from the outside and been confined in an enclosed space, and in which the working of the machinery takes place in a noiseless manner or is not observable from the outside. My present invention has the advantage of accomplishing these ends.'

Uniquely, in his seventh and penultimate British Patent – No.3942 *Improved Frictional Clutch Apparatus*, filed on 20 November 1893 and approved on 23 December – Lührig described himself simply as an 'Engineer' rather than as a 'Mining Engineer'. In that application he claimed that this improved design of the clutch mechanism was simpler to construct, less prone to wear, and much smoother in operation, and that 'it is possible with this construction to put the clutch in and out of gear so gradually as to entirely avoid all shock'.

A criticism of his earlier clutch design had been that it had a tendency to 'snatch' when engaged, and a smoothly operating clutch was essential, as the gas motor – like all internal combustion engines until 'stop-start' became the norm in recent years – was running constantly and would otherwise lack the gradual and smooth take-up of drive already available to operators of steam engines.

While the clutch seems to have operated in a highly satisfactory manner, it required frequent replacement of the wooden pads.

The tramcar described and illustrated in his final British patent incorporated a single more powerful gas engine – still of unspecified manufacture but clearly built to an Otto design and with a fundamental redesign of the layout of the vehicle. It was this design which would be seen running in Croydon in 1894.

The 'frictional clutch' patented in 1893 (British Patent No.3942). 'Between the clutch halves' he explained, 'is introduced a disc fixed on the driving shaft and having wood cheeks against which the two clutch discs are forcibly pressed by the screw spindles.'

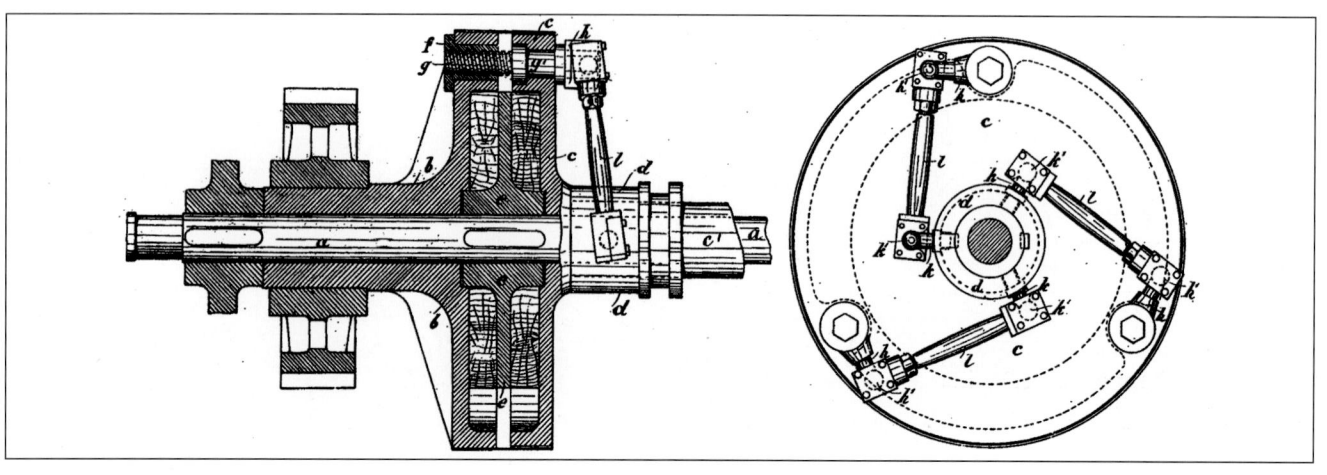

Left and below: By 1897, the Deutsche Gasbahn Gesellschaft was building its own locomotive units at its works in Dessau rather than sub-contracting the work to Van der Zypen in Cologne. Eight such units were built in 1897 and 1898 to haul trailers and former horse cars. These were fitted with larger engines. One of Lührig's locomotive cars stands to the left of a single-decked gas tram while a trailer stands behind. The company once operated thirteen powerless trailer cars. The photograph was taken on the network in Dessau north of Leipzig, probably around 1898. These two locomotives are later evolutions of the vehicle described in the 1892 patent No.19,070. They are believed to have been powered by horizontal two-cylinder engines.

Nineteenth century patents are fascinating and sometimes imprecise documents, embracing – and therefore giving legal protection to – an inventor's claim to something already in the public domain. Today, with much tighter regulations, an applicant would not be able to patent something as fundamental as a simple rethink of the layout of the tram which offered no new features, just their juxtaposition to each other – and yet that is what was afforded to Lührig under Patent No.11,506 *Improvements in Locomotive Tramway Vehicles*, submitted on 12 June 1893 and ratified just ten months later on 12 April 1894.

Almost three years elapsed between the filing of Lührig's initial patent, and the inaugural run of his first tramcars in Dresden on 27 July 1894. They operated on two of the city's tram routes – Albertplatz to Wilder Mann and Albertplatz to the Freidhof (cemetery) at St. Pauli.

The tramcars – built by Carl Stoll of 56 Leipzigerstrasse, Dresden – were fitted with 10hp engines by the company Otto had founded – Gasmotoren-Fabrik Deutz – and could carry thirty-six passengers.

Carl Stoll (1846–1907) would later enjoy widespread success for designing the overhead pick-up system for 550-volt DC electric trams which would carry his name and be widely used on many tramways.

The gas cars were, however, rather inefficient, carrying only enough gas for one round trip before having to return to the depot on Grossenhainer Strasse to refuel.

According to passengers, there was a constant smell of gas – denied by the company – causing a few concerns about safety, and those were realised when, on 10 December 1894, there was an explosion at the tramway depot which caused minor damage. The accident happened, not with the tram, but while refilling one of the gas storage tanks from the city's main gas supply.

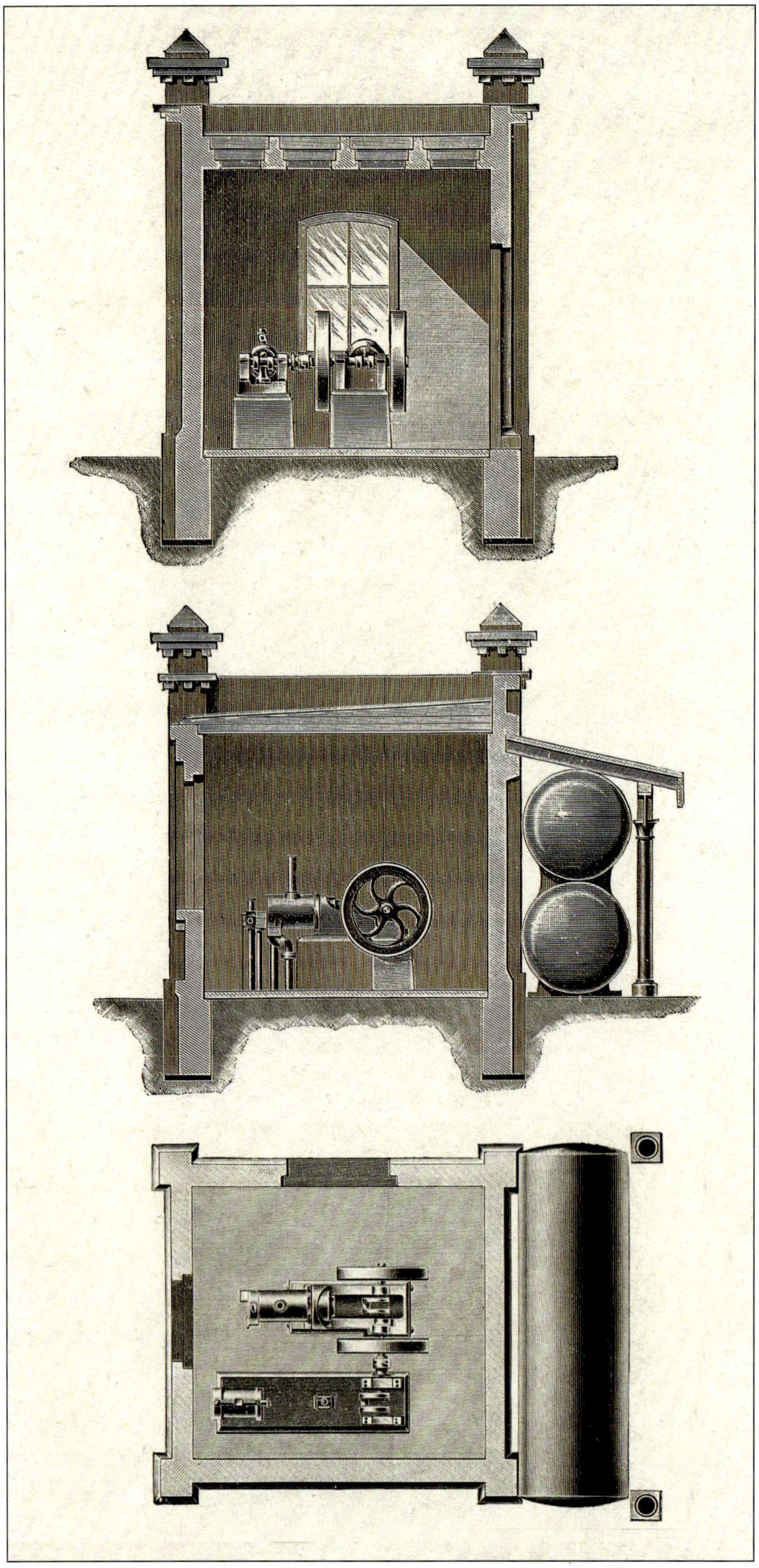

On 14 November 1894 the first of Lührig's gas trams ran on the streets of Dessau in Saxony-Anhalt, Germany and, in an article about the tramway and its vehicles, the readers of the *Scientific American* journal on 1 February 1896 were shown a single-decked vehicle, larger any of those which had been illustrated in Lührig's many patents, but was of a size and design already operating in Dresden.

The Gas Traction Company had signed a contract with the municipal authorities in Dessau to operate the system under licence, for which they would receive 20 per cent of the profits until the contact came to an end which, unrealistically, was not required to happen until 1 July 1953. In the event, the full term of the contract was not fulfilled.

Under the heading 'GAS MOTOR CARS OF THE DESSAU TRAMWAY', the *Scientific American* article gave a lot of detail about the tramcars:

> 'The first section of the gas-motor tramway of Dessau was inaugurated November 14 1894, and the second on the 16th of December of the same year. The total length of these two sections is 2½ miles.
>
> The track is of the normal gauge of 4.75 feet between the rails. The rails are the same as those used on

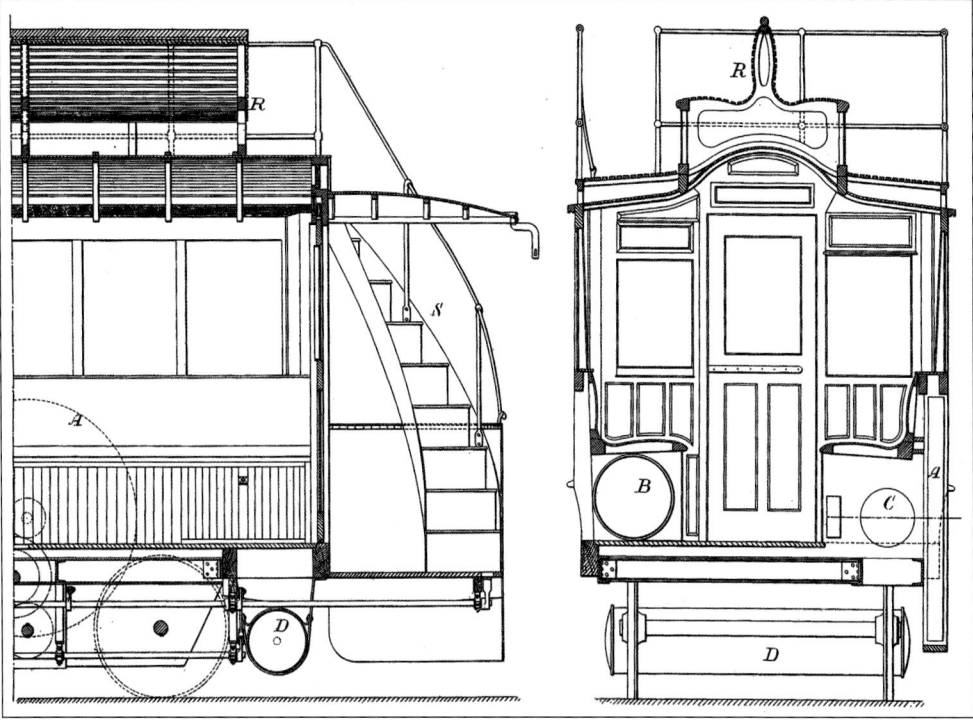

electric tramways. The maximum gradients are two-thirds of an inch to the foot, and the sharpest curves are of 40 feet radius.

The rolling stock consists of nine automobile cars of the small type of the Lührig system. The weight of each car, ready for running, is six tons. The car has a capacity for twelve passengers standing upon the platforms and for fifteen seated, say, with the conductor, for twenty-eight persons.'

The description is interesting, suggesting that readers of *Scientific American* would already know of the 'Lührig System', yet neither the Dresden tramway, nor the Croydon and Thornton Heath Tramway – the first two lines to operate gas-engined trams in 1894 – had previously been reported in the journal. The article continued:

'The car has the aspect of an ordinary street horse car. The accompanying figure gives a view of one of the sides (that on which the motor is placed), with the doors that serve for the inspection of the motor removed. The cars, with the exception of the motors, were constructed at the Van der Zypen & Charlier works, of Cologne. The motors are from the Deutz works, of the same city. The motor, which is of the Otto type, is horizontal and has two cylinders in tandem, situated under one of the rows of seats of the car. It is of an effective 7 horse power but is capable of developing ten per cent more. The transmission between the motor and the axles is so arranged as to communicate to the car a speed that may reach 7 or 9 miles an hour. The car easily ascends gradients in hauling another car full of passengers and not provided with a motor. Since the opening up of the line, the company has purchased freight cars and trailing cars.

The car carries three gas reservoirs, two of which are placed under the platforms and one under the row of seats opposite that under which the motor is situated. Their total capacity is 26 cubic feet, which suffices for a round trip.'

Opposite left: Three illustrations of a design for a building to house the compressor engine and gas storage tanks for the Lührig system. These illustrations come from *De Gasmootortram volgens het Systeem Lührig* published in Amsterdam in 1896 by 'The Gas Traction Company Departm. Netherlands and its Colonies'. This sixty-three-page promotional booklet was intended to prepare the citizens of Amsterdam, The Hague and Maastricht by extolling the virtues and benefits of gas traction over the alternatives in advance of the planned trials of the system.

Opposite right: One of the Dessau compressor stations as illustrated in an article on the Dessau tramway in Schilling's *Journal für Gasbeleuchtung*, 1895. The flywheel of the gas compressor engine can be seen through the window.

Above: The double-decked tram design described in Patent No.11,506 of 1893, Lührig's eighth and final tramcar patent filed in Britain. Beneath the left-hand seat is one of the rubber gas bags, while under the right-hand seat sits the two-cylinder horizontally opposed Otto-cycle engine. The first version of this design was possibly one of the cars tried out on the Croydon and Thornton Heath Tramway in 1894. A modification of this vehicle – with six windows each side and reversible seating on the top deck – was later used on the Blackpool, St. Anne's and Lytham Tramway from 1896.

In Lührig's earliest patents in Germany, the United Kingdom and the United States, the cars illustrated all had four or six windows each side, rather than the five shown in this illustration of a 'Gas Motor Car of the Dessau Tramway' published in *Scientific American* on 1 February 1896. This size of car was built in early 1894 for use in Dresden, later being sold to Dessau before being tested on the Croydon and Thornton Heath Tramway in May 1894. It was later shipped to France and run briefly on the streets of Paris along with one of the small double-decked cars of the type later used in Blackpool and Neath. This car carries the legend 'The Gas Traction Company Ltd. London' under the 'Tramway a Gaz' banner. A profile diagram of a five-window car appeared in *Cassier's Magazine* in June 1895 although the engravings of Dessau's trams which accompanied the article all had four windows each side. The 'life protectors' – more commonly referred to as 'cow-catchers' were added at the insistence of Major-General Charles Scrope Hutchinson CB, Chief Inspecting Officer for Railways from 1892 to 1895.

This photograph shows the opposite – non-engine – side of the Dresden-style tramcar from which the illustration in *Scientific American* was drawn. It is seen here in a Paris park in 1895. When tested in Paris, according to the French writer M. Lavezzari, its limited passenger capacity was considered unsatisfactory and it was returned to England.

The choice of Van der Zypen und Charlier to build the cars was an obvious one. Coachbuilder Ferdinand van der Zypen from Liège and the freight handler Albert Charlier had opened their first wagon factory in 1846 west of Deutz-Mülheimer Strasse in Cologne, and they were acquainted with Eugen Langen, Nicolaus Otto's partner in the Deutz engine works, and co-inventor of the gas engine. They would later jointly develop overhead railways in several German cities.

Two contemporary photographs of the larger capacity double-decked Blackpool tram which was repainted and trialled on the streets of Paris after the single-decked vehicle was rejected. It also carries the name 'The Gas Traction Company Ltd. London' under the 'Tramway a Gaz' banner.

Dresden tram No.127, decked out in flowers and bunting, perhaps for its inaugural run. The numbering does not imply the city had 127 gas trams – it actually had five – and the numbering sequence is believed to have followed on from the city's fleet of horse cars. Only four of Dresden's five cars – which were operated by the Gas Traction Company – seem to have been transferred to Dessau in 1896. The fifth vehicle is thought to have been the one repainted and sent on trial to Paris in 1895. According to M. Lavezzari, it was then shipped back to Britain where he saw it at the Ashbury Carriage Company's works in Manchester.

The article described the charging stations where the 'illuminating gas' – so called due to its regular use for building and street lighting – was compressed into tanks to await the trams.

The compressor was powered by an eight horsepower Otto gas motor and the tanks contained sufficient gas to charge two tramcars, so it was only necessary to run the compressors about three hours per day to keep the system fully charged. About 10 per cent of all the gas burned was used in powering the compressors.

But while Lührig and the Gas Traction Company were making progress in Europe, The Connelly Gas Motor Company was also developing a vehicle at its works in New Castle, Pennsylvania. They clearly had big plans for it.

Their compact gas-engined locomotive, designed to haul former horse tramcars, was fitted with a vertical two-cylinder gas engine, and the company envisaged huge sales for it across the world. The front page of the *New Castle News* carried the story on Monday 14 November 1892:

> 'The Connelly Gas Engine Works are preparing to make improvements on their present plant to the extent of $15,000. A new building is being put up 50x100 feet, which will reach in the rear of the present plant to the railroad. The old machine shop will be utilised also, and employment for 150 men will be given. This will mean that one gas motor will be turned out daily, and that New Castle will have a hand in supplying the demand for motors.
>
> Mr. Connelly is interested in another company which has a contract to produce 500 of these motors for the city of Chicago, and yet another company which will make an equal number for London. The motive power is the cheapest in the world, and recommends itself wherever tried.'

In the event, the widespread use of the Connelly locomotive never materialised, although a modified version of it was built and briefly run in Britain.

Back in Europe, the Gas Traction Company was doing its best to encourage tramways to try out their vehicles, but with quite limited success.

In 1895, a proposal was tabled to build and operate a gas-engine tramway in the south-western suburbs of Berlin, between the railway station at Zehlendorf and Lichterfelde. The Gas Traction Company was to build and operate the line but it does not seem to have been constructed. Also in 1895, a committee was set up to investigate the best tramway system for the city as a whole and gas traction was one of those being considered, although Berlin already had electric trams – the Gross-Lichterfelde Tramway had been constructed by Siemens and opened in 1881 and is said to have been the first in the world.

That committee recommended comparison trials along a designated stretch of line hitherto served by horse-drawn trams.

From the several companies which offered to demonstrate their trams, three radically different systems were selected for the trials – electric vehicles from Siemens, French Serpollet steam trams, and Lührig gas cars.

It would seem that even then, there were admitted concerns about the power of the gas tram engines, for when one of the Dessau team went to look over the planned route, he concluded that it would considerably disadvantage the gas tram.

The Gas Traction Company team made the decision to withdraw from the trials, promising to demonstrate their car elsewhere at a later date – presumably on a route the topography of which would not draw attention to the power issues of which they were already aware. Their decision effectively sealed the fate of the gas tram as far as Berlin was concerned and the city went electric.

Elsewhere, however, the system found a more encouraging response in the Netherlands, with gas vehicles being given lengthy trials in Amsterdam, The Hague and Maastricht.

Attention was drawn to these tests in the *Westminster Gazette* on 22 June 1899 in a letter from Thomas Hersey – a strong advocate of gas traction and an associate of the Gas Traction Company – who was clearly hoping to persuade London's tramways to explore the system:

> 'One of the gas-driven cars, similar to those running from Blackpool to Lytham, has been working on the public tramways at The Hague for six months without stopping for a single day, with the result that the Committee of The Hague Municipality reported in favour of this system of mechanical traction, as against horses or electricity, and this decision was published in the Amsterdam and Hague Journals.
>
> Without exception, all the engineers that I am acquainted with who have examined these gas motors have reported favourably on the prospects for gas traction.'

During the tests of that car on Dutch metals, it reportedly ran for more than twenty-one miles on a single charge of gas, compared with the ten to twelve miles which earlier cars had achieved.

Back in 1896, however, a paper was read to the Société des Ingenieurs Civil de France by M. A. Lavezzari and published in the Société's *Mémoires et Compte Rendus des Travaux* in 1896, describing progress to date in great detail:

> 'The idea of using gas as the driving force of tramcars is already old, and several attempts have been made by various inventors. But the only one, to my knowledge,

In the autumn of 1900, this gas car is said to have briefly been trialled on one of Amsterdam's tramways. It is clearly of a different design to other similarly-sized Lührig gas cars, with the engine and flywheel set much lower in the frame. The test runs were carried out over several weeks, but only at night after the normal service had ended for the day. It was not considered a success.

which has been used under real operating conditions, which does not mean that it is the only practical one, is the Lührig system, from Dresden, which has since become that of the Gas Traction Company. The first experiments took place in 1891, but it was not until July 1894 that a regular service was organised in Dresden.

The results were such that in December of the same year, a company was formed in Dessau for the large-scale exploitation of this system. Quite recently a new line has been put into operation in England at Blackpool, and finally the Parisian gas company is currently testing.'

The Paris tests were given a complete and detailed section to themselves in Lavezzari's presentation. The first tram to which he referred, a single-decker (*see page 78*), was in all probability the Dresden car which had been used for the second Croydon tests.

'Our large gas company has not remained indifferent to the development of traction by means of gas engines; so it came to an agreement with the Gas Traction Company to import a test car; a first was sent to Paris, but it was too small to give interesting results. It was returned to England after a few runs and another, larger and with an upper deck, has just undergone a series of trials here. The first experiments took place inside the Saint-Denis depot; then it made several trips between the Porte de la Chapelle and the Place aux Gueldres, in Saint-Denis.'

He included a great deal of fascinating data from those test runs, but did not indicate the duration of the trial or the mileage covered by the vehicles. He did, however, feel the need to acknowledge the source of his information:

> 'I owe this information to the kindness of Mr. Lévy and Mr. Bertrand, Engineers of the Compagnie Parisienne, to whom I address my thanks. Their high competence is the best guarantee of the integrity of the tests and the accuracy of the published results.'

While Lührig certainly saw a great future for his system, envisaging his trams operating worldwide, his plans from the outset involved selling his patent rights country by country to other companies.

By the time of the *Scientific American* article, the Dessau line had been extended by a further 1¼ miles, and the 42,500 residents of Dessau had, the magazine reported, 'developed a taste for riding'.

To deal with that increased demand, the number of tramcars was increased from nine to thirteen. The four new cars, built at Deutsche Gasbahn Gesselschaft's own works in Dresden, had 10hp Otto-cycle engines fitted, presumably of a larger capacity, although that is not stated. 'So the success of the enterprise has exceeded all expectations' the magazine reported, concluding 'It is evident from these data that the tramway company is satisfied with its system of propulsion.'

A further four cars were added to the fleet after the Dresden tramway system started withdrawing its gas vehicles on 15 March 1896 – the last gas car ran just before the end of the year – resulting in Dessau having a fleet totalling seventeen powered tramcars together with eight locomotive units which were used to haul an unspecified number of former horse trams. The fifth Dresden gas-engined car was briefly moved to Paris as a demonstration vehicle.

The 6-wheeled Gilliéron & Amrein power cars had the engine at the front, driving a four-wheel front bogie. The driver therefore had to operate the tram in a confined space with the large flywheel (visible behind him in the photograph) and engine running alongside him. Having the engine and driving controls only at one end imposed the additional requirement of turntables at either end of the line to reverse the direction of travel.

Carl Lührig was not alone in seeing the potential of gas-powered vehicles. Indeed the idea can be traced back to 1879 at the latest when Francis William Crossley – one of the Crossley Brothers who would do much to popularise the gas engine – and Henry Percy Holt started discussing using a gas-engine to haul former horse trams. That is discussed in the next chapter.

But while Lührig was filing patents, others were also developing the idea, and while the Croydon trials were taking place – first with the Connelly locomotive and then the Lührig cars, a gas tram service opened on 20 January 1894 between Neuchâtel and Saint-Blaise in Switzerland.

The tramway had two gas-powered tramcars and two unpowered trailers built by Gilliéron & Amrein of Vevey on Lake Geneva, and the service had been intended to start in October 1893.

Late delivery of the powered cars and even later delivery of the trailers meant the service could not start until three months later than planned.

Even then, it was something of a disaster. The 8hp 'Gnom' engines from Motorenfabrik Oberursel proved hugely under-powered and were quickly replaced by more powerful 15hp Benz engines, but even they proved inadequate when pulling trailers.

Neuchâtel tram No.1 taking on gas, probably at Gilliéron & Amrein's works in Vevey. The part-built chassis and engine of the second power car can be seen to the right. This picture probably predates the start of scheduled services in January 1894. Note the removable side screens either side of the driving compartment.

THE FIRST GAS TRAMCARS • 87

The Neuchâtel powered cars were intended to work either singly or with a trailer depending on demand. Each vehicle had relatively low capacity – twelve seated and eight standing on the rear platform – many fewer than Lührig was envisaging for his vehicles. The trailers were planned as an essential part of making the route commercially viable. However this photograph is believed to show the two powered cars, with the first about to depart from the depot.

The vehicles were very different to Lührig's design. The engine was at the front in the driving compartment, and despite the fact that they used larger engines, they suffered the same power issues many operators of Lührig's cars reported.

Faced with the tramway company refusing to accept the vehicles, Gilliéron & Amrein even offered to replace the gas engines with petrol engines, but that too was rejected in favour of a return to horse-power.

Another suggestion from the builders – that the trailer cars alone be hauled by horses – also proved unworkable due to their weight, resulting in a quite substantial investment in gas power being completely written off.

Elsewhere, the authorities in Jelenia Góra – then part of the German Empire but now in Lower Silesia in south-western Poland – chose Lührig's small gas cars as the for their new tramway, opened in 1897.

The choice of gas was simply for expediency, as the town did not yet have an electricity generating station.

By 1900, however, the authorities' attitude towards gas trams had changed – again due to their lack of power. In the meantime a power station had been built and the erection of overhead wires heralded the introduction of an electric system.

Just a year later, on 24 March 1901, the last gas tram ran in Dessau. Lührig's gas system had lasted only seven years in his native Germany – far short of the sixty his contract had envisaged, and nowhere near as long as it would elsewhere.

After visiting Dessau in February 1895 Jos Bauduin introduced three small gas cars – similar to those in Dessau – on to the streets of Maastricht in the Netherlands in 1896, on a route, between Wyck Station and Vrijthof.

The tramway included a slight incline at either end of the Sint Servaasbrug – the Saint Servatius bridge over the River Maas or Meuse, which the small Lührig cars apparently had insufficient power to cope with when carrying a full load.

Right and below: Lührig tramcars started operating in Jelenia Góra in the Hirschenberg Valley – then part of the German Empire but now in south-western Poland – in 1897. The fleet size is not known, but from the car illustrated right, it must have numbered at least ten. The number on the car in the stereoscopic (3D) view, below, cannot be seen. Gas trams were already being augmented by electric traction with overhead wires by 1900, the gas fleet being withdrawn shortly thereafter.

It is said that when they approached either side of the bridge, the twenty-one unfortunate passengers sometimes had to get off and push – despite all having paid their fares. The gas cars were withdrawn in 1903, the tramway reverting to horses for a period before electrification.

A tram of similar design to the Maastricht cars was also briefly trialled on the streets of Amsterdam – part of a proposal to build a gas tramway between Amsterdam and Haarlem – but the project was abandoned, Amsterdam being one of a number of towns and cities that deemed the vehicles unsuitable, expressing concerns both about their capacity and about the safety of the compressed gas tanks.

Around the same time as the Maastricht line opened, the Eerste Nederlandsche Gastractie Maatschappij – The First Gas Traction Company of the Netherlands, which is presumed to have acquired the rights from the British Gas Traction Company's Dutch division – supplied two gas-engine trams to the Westlandsche Stoomtramweg Maatschappij (The Westland Steam Tramway Company) based in Loosduinen, a beach suburb to the south-east of the Hague, to operate on their service into the city centre. The service operated for several months – apparently successfully – but exactly why and when the gas cars were withdrawn has not been ascertained.

As far as Lührig is concerned, a question remains as to what triggered a mining engineer's interest in gas trams, and why it took so long to bring his huge investment in this innovative mode of transport into public service.

An account of gas tram development had been written in 1893 by A. Kemper in the *Journal für Gasbeleuchtung* (the Journal for Gas Lighting). It was translated and re-published in *Tramways – Their Construction and Working* by D. Kinnear Clark published in 1894. Kemper's assessment of the relative costs of the tramway systems then in use is enlightening:

'Only twenty-six years have elapsed since the first horse-tramway was established in Germany, but in 1889, in about fifty towns, there were 838 miles in use, while in the United States much greater progress has been made.'

There then followed a brief summary of the various systems available – horse, steam, cable, battery, trams with benzene or naphtha motors, and the experimental system then being trialled in New York City using carbonic acid as a fuel:

Postcards of Lührig cars in operation are rare, but an interesting card, c.1897, produced in the few years gas trams operated in the Hirschenberg Valley celebrates not their virtues but their shortcomings. On it, pedestrians are seen rushing out of the way as a fully-laden tram descends a hill, while others are seen pushing their tram uphill. Similar criticisms were made of the Maastricht cars when crossing the Sint Servaasbrug – also known as the Massbrug (*see overleaf*).

Right and below: Postcards from 1902–03 of Maastricht's Gastram No.2 making its way across the Maasbrug. Several other postcards of gas trams crossing the bridge were taken from the same viewpoint.

Gastram No.2 making its way through the centre of Maastricht. Beyond it is one of the horse-drawn tramcars. Three gas cars were operated between 1896 and 1903. Apart from what appears to be smaller cooling water storage facilities on the roof, these cars are similar in many respects to those operating in Jelenia Góra. The cars carried the numbers 1 to 3 at the front and back, and a two-tone livery on the sides.

THE FIRST GAS TRAMCARS • 91

Maastricht Gastram No.1 on its way to the depot at Lindenkruis.

'For working tramcars with gas the receivers are six to ten in number, having a total capacity of 44 to 88 cubic feet, and are placed beneath the floor. The gas supply is taken from the mains, only a supply-station and a small compressing engine being needed; one or more such stations being provided, according to the length of the line, each station costing £400 to £600. A plan for a compressor station is given, showing

Maastricht's Gastram No.1 part way across the Massbrug – the 190 metres long, thirteenth century, bridge which crosses the River Maas or Meuse. Even the gentle incline at either end of the bridge proved too steep for a fully laden gas tram, with passengers having to get off to lighten the load.

Maastricht Gastram No.1 at the terminus outside Meerssen railway station.

an 8-H.P. gas-engine with a compressor capable of compressing 2,130 cubic feet of gas per hour. The gas, taken from the main, is passed through a meter to two storage vessels, having a united capacity of nearly 400 cubic feet. At a pressure of 8 atmospheres they will supply the reservoirs of two cars at 6 atmospheres. The gas required for compressing purposes is about 8 per cent. of the gas to be compressed. Two types of cars for gas-tramways have been introduced in Germany. Those of Messrs. Guilliéron and Amrein are used on the line between Neuchatel and St. Blaize. There is an 8-H.P. gas-engine on the outside platform of the car. The gas receivers are sufficient for the double journey of over 3 miles each way, the consumption being estimated at 34 cubic feet per car per mile. An empty car weighs 6 tons, will carry 20 passengers, and costs £750.

The other type is that of Lührig of Dresden. Each car is driven by two 7-H.P. gas-engines, fixed under the seats, with the flywheels behind the backs of the seats. The engines can be worked together or separately, and arrangements are made for three rates of speed, 150 revolutions per minute when engines are running free, 200 revolutions for low and 240 for high speed.'

Unless Dresden had replaced its entire fleet within the first two years of inaugurating the service, that account is at odds with Lavezzari's description in his paper to the Société des Ingenieurs Civil de France in 1896, where he describes the Dresden cars as being single- rather than twin-engined, as were those also in use by that time in Dessau, Blackpool and Neath.

'The engine is of the Otto type, four-stroke, with two horizontal cylinders facing each other; the two cranks are an extension of one another so that there is an explosion per revolution of the flywheel. The two distributions are made by conjugate valves arranged so that the admission of gas in one cylinder coincides with the compression of the other. The engine is placed under the bench on one side of the car, and the flywheel is housed within the side wall, which is made up of a double partition on this side.'

However, back to Kemper, who, in his 1893 description of the system, continued:

> 'If the car remains only a short time at the stopping or end stations, the engines run alone, to avoid re-starting. Ignition is effected by means of a small electro-magnetic lighter. The condensing tanks are on the roof of the car, with automatic circulation. There are three shafts, one with toothed wheels worked direct from the engines and wheels for varying the speed. The large cars weigh when empty 7½ tons, and with 29 persons 9½ tons, and will ascend an incline of 1 in 23 at a moderate speed. A smaller car, better suited to steep gradients, has been constructed, and is worked with a 10-H.P. engine. This car weighs 4½ tons, and holds 22 persons, and would ascend an incline of 1 in 15 at a speed of over 3½ miles per hour.'

Thus the rationale for the two sizes of cars operated in Dresden and Dessau is explained – something Lührig had not mentioned in his patents. There is no definition of what is considered 'a moderate speed' for such a modest gradient.

> 'The cost of a large car is estimated at £900, and of the smaller one at £700. The gas consumption, with 10 to 12 passengers with the large car, was found to be 34.7 to 37 cubic feet per car per mile. For the cost of construction it is estimated that, for a line about five miles long, with cars running every five minutes, which would require 20 cars, and with an average working day of 14 hours, the cost of a gas-worked tram line, including rails, cars, buildings, &c., would be £6,040 per mile. Under similar conditions an electric tram is estimated to cost £7,648 per mile, and a horse-tram £5,636. The working expenses with gas at 3s. 5d. per 1,000 cubic feet are estimated at about 3 pence per car per mile, for horse-trams, with 1-horse cars at 4.25 to 5.4 pence per car per mile, and for electric trams at 3.86 pence per car per mile; and it is shown that, with similar traffic conditions, a gas-tram might be expect to give a return of 6½ per cent. on the capital invested, while an electric tram would barely cover the cost of working.'

That report raises some questions for which answers have proved elusive. Kemper notes that 'Two types of cars for gas-tramways have been introduced in Germany. Those of Messrs. Gilliéron and Amrein are used on the line between Neuchâtel and St. Blaize' – but both Neuchâtel and St. Blaise are in Switzerland, not Germany. Was the Gilliéron and Amrein system used somewhere in Germany? If so, no evidence has yet been found.

Records of Carl Lührig's first successful run of his tramcar in Dresden have not survived the turmoil through which that city has passed in the last century, but his first patent must be assumed to mark the point in his work where he had something original to protect.

That would place his first trial in 1891 – a date effectively confirmed by Lavezzari in his paper to the Société des Ingenieurs Civil de France.

But at the time Kemper was writing about the Dresden and Dessau trams in late 1893, it is clear that it was the twin-engined cars built to Lührig's September 1892 patent which he had been able to see in action.

Seven years before Lührig was granted his first British patent, and several thousand miles away from his Dresden base, another gas tram set off on its first passenger-carrying journey in Australia – between the northern Melbourne suburbs of Clifton Hill and Alphington.

The line on which it ran had been built by Victorian Railways a few years earlier but not yet brought into use as a suburban railway. Thus it was ideally suited to the experimental operation of a gas tram service.

The date was 1 February 1886, and that first run is recorded as having left Clifton Hill, on time, at 7.30am. For the next two years, a single tramcar operated six days a week, with eight round trips Mondays–Fridays and a ninth on Saturdays. The journey from Clifton Hill to Alphington – just under two and a half miles – took approximately fifteen minutes, at an average speed – quite reasonable for a tram at that time – of just over nine miles per hour.

The tram was introduced by two engineers, Benjamin Barnes and John Danks. Barnes was involved with Victorian Railways while Danks came from a manufacturing background and described himself as a 'brass founder'.

Their interest in exploring gas traction was in response to the capital cost of installing Melbourne's cable tram network – the city's first cable route had been opened in 1882 – and their first experiments were carried out during either late 1884 or early 1885. They used a six-wheel vehicle of their own design and construction fitted with a single-cylinder engine, fuelled from two large India-rubber gas containers underneath the bench seats on either side of the car. The original engine, at just 3.5hp, proved to be inadequate and was replaced by a 6hp version.

By February 1885 they had built a prototype four-wheel tramcar with an integral gas engine mounted horizontally at one end of the vehicle. For that design they had been granted a patent both from the Victoria and South Australia Patent Offices on 11 November 1884 as Patents No.3,882 and No.554. The following year, the same specification was contained within US Patent No.320,634, issued on 23 June 1885.

It is important to understand at this stage that the prototype vehicle was never intended to run on the recently built cable tramlines, the first routes of which had been opened in 1884 and 1885.

This was purely an experimental vehicle and it had been built to the 5ft 3ins gauge of the Victorian Railways – rather than the standard gauge of the recently introduced cable trams – in order to take advantage of the availability of the Clifton Hill to Alphington track as a proving ground.

That entirely pragmatic decision taken by the two inventors has led, over the 140 years since, to some very confused reporting on the design and purpose of the 'tramcar'.

The first run of their working vehicle was reported in the Melbourne newspaper, *The Age*, on 3 February 1885 headed 'An Ingenious Motor':

'In view of the inauguration of a system of tramways in the city of Melbourne and suburbs and the great importance of employing silent motive power in busy thoroughfares, considerable importance was attached to a trial of a new motor for tram cars made on Friday morning at the works of the well known firm of brass founders, Messrs. John Danks and Co., South Melbourne. A large number of gentlemen, amongst whom were several of the leading merchants and public men of Melbourne, were present, and after witnessing the trial expressed the opinion that the inventors had solved the difficult problem of obtaining a powerful, silent and thoroughly efficient motive power for drawing tramcars through the crowded streets of a city. The new principle which has been worked out and applied by Mr. Danks and Mr. Barnes, the latter gentleman for many years connected with the Victorian Railway department, is simplicity itself, and the wonder is no one ever thought of it before. It consists of an Otto silent gas engine, of five horse power, fitted into the fore compartment of a tramcar. On the flywheel shaft two friction wheels about 18 inches in diameter are keyed and work into two other friction wheels keyed on to a counter shaft, which works a pitch chain. The latter passes around a pitch chain disc on one of the axles of the tramcar, so that when the chain revolves it causes the axle to move at the same time. In order to simplify the description it may be stated that the friction gear is exactly the same as that used in winding machinery for mining plants, and that the pitch chain is an exact counterpart of the chain employed in propelling a tricycle. The engine once started need never be stopped for a whole day. All that is required to stop and start the tram car is to move the two levers attached to the friction wheels.

At the trial which took place on Friday the engine was fitted in a tram car constructed specially for the occasion. The car was large enough to carry thirty passengers, and although in an incomplete state the visitors had an excellent opportunity of judging of the superior method of travelling by means of tram cars compared with an omnibus or cab. The gauge of the rails is the same as that of the Victorian Railways,

The key diagram which accompanied Danks and Barnes' South Australia Patent No.554, granted on 11 November 1884, the same day as their Victoria Patent No.3,882 was granted. The horizontally mounted engine was linked by friction clutches and chain drive to the wheels. The specification was replicated in US Patent No.320,634, granted to them on 23 June 1885.

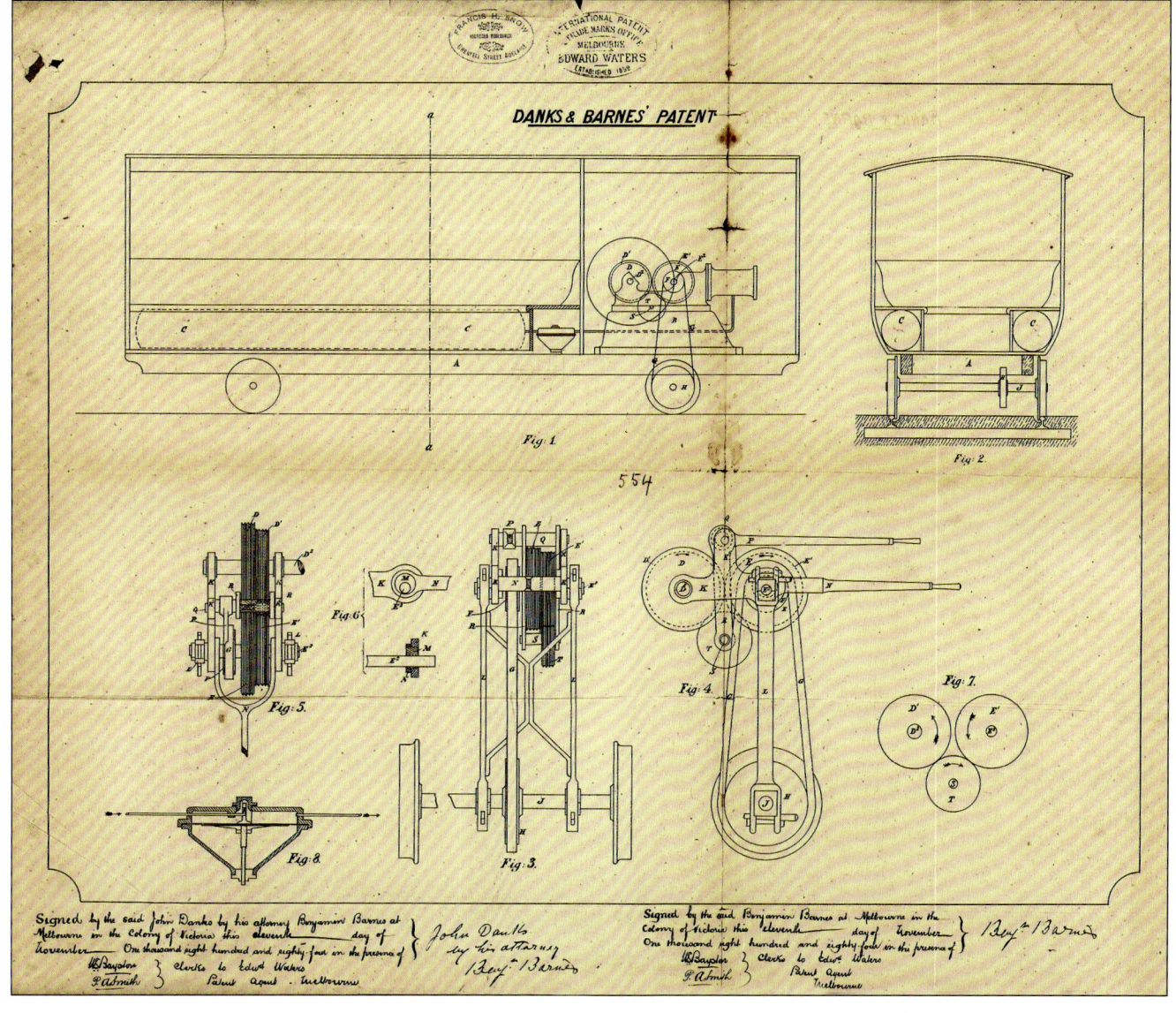

The final page of Danks' and Barnes' patent claims patent protection for features which would also be found in patents registered by inventors in Germany, Switzerland and the United States both before them and after them – namely the gas engine contained within the car, gas storage containers beneath the seats and friction clutches to enable the tram to be stopped while the engine was still running. All of these had already been patented in Germany by Conrad Krauss and in Britain by J. and J. Quick, amongst others.

the object of the designers being to run the car as an experiment on the completed portion of the Alphington line. An application has been made to the Government for permission to make the experiment, and there is no doubt it will be granted. All wheels, of which there are six, are braked on each side. The brake is a powerful one, and is applied by the driver pressing his foot on a plate projecting through the floor of the carriage. In order to obtain a sufficient supply of gas to last for four hours' travelling at a maximum rate of 6 miles an hour, several wrought iron cylinders are fixed under the seats of the car in such a manner that they cannot he seen.

Into these cylinders the gas is forced up to a pressure of 100lb to the square inch. As a gas engine will not work unless the pressure is brought down to $1/10$th of an inch water pressure, similar to that of the gas mains, two reducing valves are used to bring the pressure down to the proper degree. When the engine was started on Friday it worked exceedingly well.

The wheels and axles flow round with extraordinary rapidity. The noise of the explosions of gas were very faint as they passed through two air chambers. As soon as the car is completed. which will be in the course of another week, it will be removed and tried out on the Alphington line. The experiment will be watched with considerable interest, for should it prove a success it will effect a complete revolution in the method of propelling tramcars. The motive power can be supplied at less than half the cost of the cable system. This, together with the fact that the cost of the gas consumed in one hour is estimated to amount to 6d, should command the earnest attention of the Melbourne Tramways Trust, and the bodies interested in the construction of tramways in Melbourne and suburbs.'

Testing the vehicle under operational conditions for several months was not without incident – just four months into the test programme, the *Kyneton Observer* reported on 19 May 1885 on what at least was an act of wanton vandalism, or at worst sabotage – no later newspaper reports have been located which might answer that.

'DIABOLICAL ATTEMPT TO UPSET A TRAMCAR,
One of the most fiendish and diabolical attempts to wreck a tram car we have had to record was made on Wednesday afternoon on the Alphington railway. A party of gentlemen were travelling from Clifton Hill to Holly Bank Station on a gas-motor tramcar for the purpose of testing its capabilities and speed. The journey downward was accomplished in safety. On the return trip betwixt the Fairfield Park and racecourse, there is a sudden curve. On nearing this point one of the persons in charge observed several stones on the rails. The alarm was given to the driver who instantly stopped the carriage, and consequently averted what might have proved a serious loss of life. Only about seven minutes had elapsed since the carriage had passed over this spot. There were four persons within a few yards of the scene which rendered it utterly impossible for the fiendish deed to have been committed without their knowledge or participation. Yet these persons upon being interviewed absolutely refused to give any information about the perpetrator of this diabolical act. It is to be hoped the police may be enabled to track the criminal offender and speedily bring him to justice.'

Despite that mishap, testing and modification of the car continued and in January the following year Danks and Barnes were sufficiently confident of their vehicle to get permission to operate a scheduled service on the Alphington line, starting on 1 February 1886.

When the lease expired in 1888, control of the line reverted to Victorian Railways. The route was then extended, and operated thereafter as part of their suburban network using conventional steam locomotives hauling conventional carriages, and of course still using their broad gauge. The standard-gauge street tramcar which Danks and Barnes had planned seems never to have been built.

Just three months after their service carried its first fare-paying passengers in 1886, details of another gas tram were submitted to the US Patent Office on 28 April 1886 by Jay Noble of St. Louis, Missouri.

Noble worked for the long-established M. M. Buck Manufacturing Company of St. Louis who sold everything from padlocks and lanterns to railway equipment. His engine was described by the *San Francisco Bulletin* four months before the patent application on 29 December 1885:

'The motor was fixed in a car exactly similar in size, shape, and capacity to a cable grip car, and when the whole is perfected it is claimed that there will be 20 seats for passengers. It will differ in appearance from a grip car only in the fact that the engineer will stand in front instead of in the centre. The motive-power is gas, with

98 • THE GAS TRAMCAR

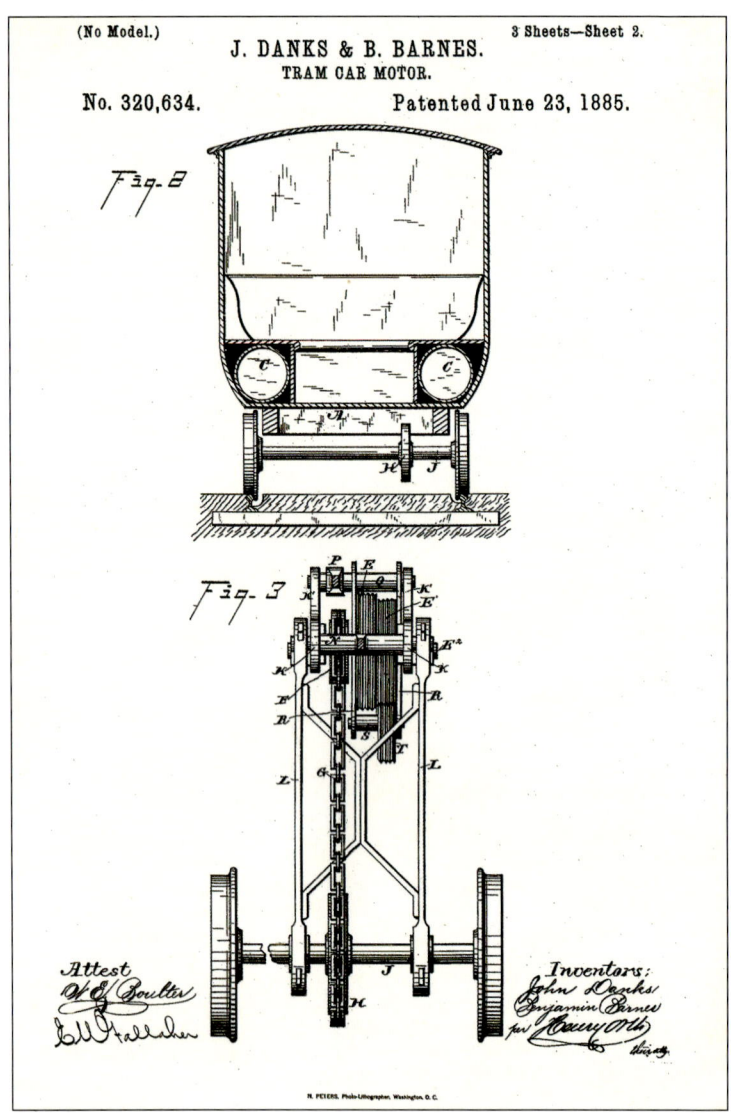

which the reservoir is to be charged once a day at power stations by means of a rubber hose. Absolute silence in working is secured by leather friction reducers, and a great feature is the production and storage of electricity while the car is in motion. This is used for lighting up the tram, and also for driving the engine on steep grades and effecting a start, in the course of the trial many interesting and novel features were disclosed. Thus the same lever is used for starting ahead and reversing, and also for releasing the driving-gear altogether.

When the car is stopped the machinery remains in motion, and a great difficulty, that of starting on a grade, is thus got over. Wherever a car can be held by a brake it can be started, claims the inventor, and any possible sliding of wheels can be got over by means of sand-boxes. The speed can be varied to an almost unlimited extent, and the car was propelled at the trial at a snail's pace to show how progress could be made around dangerous corners and through crowded streets. The break-power is exceptional. As in a cable tram, all the wheels on both cars can be locked instantaneously, and the ease with which the machinery can be reversed adds a still further safeguard in the event of a sudden stop being rendered necessary. Any summer car can be converted into a Noble gas-electric motor car without great difficulty.

Above and right: Two of the illustrations which accompanied the US patent for Danks' and Barnes' tramcar, (No.320,634), making it clearer than the Australian Patent that the linkage from the engine to the front axle was by chain drive.

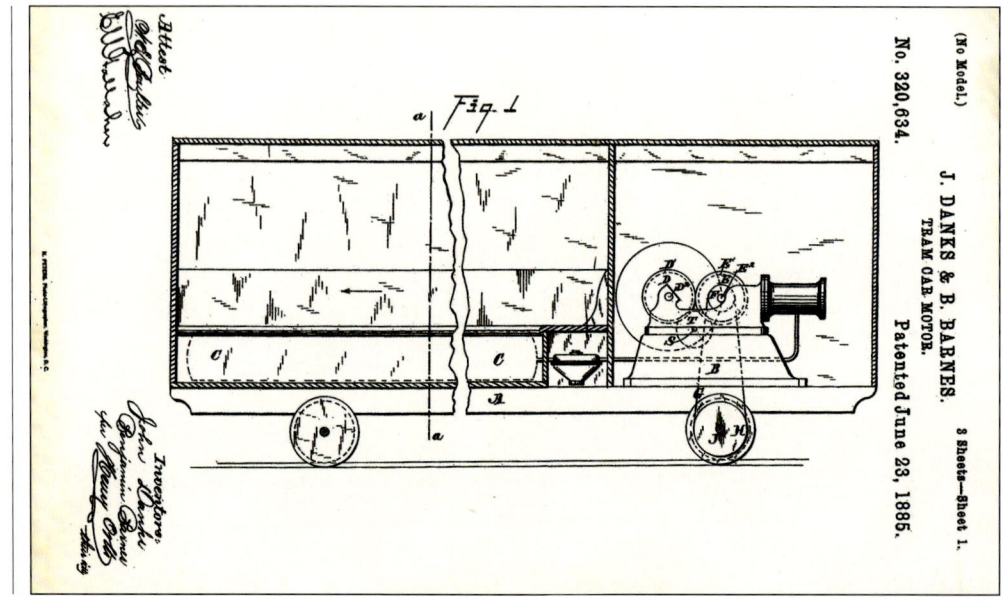

As to the prime cost of the motor and car Mr. Noble says it will not greatly exceed that of a grip-car, and he further states that 1,000 cubic feet of gas per day will more than suffice to run a tram of two cars.'

Some of those involved in the development of the tramway had already baulked at the huge infrastructure costs which would have to be met if installation of a cable system proceeded, and had great hopes for the gas alternative. The system in St. Louis was known as the 'Street Railway' and Noble's engine was described in the *St. Louis Post-Dispatch* on 22 June 1886. The novelty of an internal combustion engine was such that it still needed a fairly basic explanation if the newspaper's readership was going to get an idea of how it all worked:

'The Street Railroad Magnates, outside of the cable men, are investigating the subject of motors and have hit on one which promises to be a success and to be a vast improvement in speed, in lack of weight, and in economy over the cable system. It is the motor patented by Jay Noble, late of M. M. Buck and Co. It consists of a gas engine in which the motive power is given by a series of gas explosions caused by an electric spark and managed by a simple lever capable of reversal and running either way. The gas can be generated by the heat of the explosion or can be furnished by small tanks. The motor has stood every test and a complete outfit is now being constructed and will be finished next month.'

According to the report, the engine produced 10hp and the funding for developing the system was being raised by local tramway owners who had taken a 50 per cent interest in the vehicle 'and will adopt the motor on their lines when perfect success has been generated.' The 'perfect success' anticipated by the newspaper columnist, the project's backers and by Jay Noble himself was never achieved – the car was found to be sadly lacking in power – and the project was abandoned. The same newspaper, the *St. Louis Post-Dispatch* announced on 13 January 1887 that:

'The Noble motor was not a success. It was tried on an experimental track and whatever its defects might be the gentlemen interested kept to themselves. They have decided, however, that it will not do.'

Amongst the features of Noble's design were two design features which were intended to minimise the noise of the gas motor and limit or eliminate any pollution from it. The motor was to be fitted with 'mufflers' and the exhaust gases were to be discharged into a water tank carried on the vehicle – both things essential as the design was intended to be fitted to 'summer cars' which were open-sided vehicles.

Despite the decision not to proceed with the engine in 1887, that did not herald the end of the road for gas trams in St. Louis. Three years later, on 8 June 1890, the *St. Louis Post-Dispatch* announced that another gas-fuelled tramway was under consideration. This time it was The Cross Town Electric Railway Company which floated the idea.

The newspaper did not say whose design of engine was to be used to drive the trams, but it was probably John S. Connelly's (*see pages 115–117*).

'This company if granted a franchise will be the first to use the gas motor for street railroad transit in St. Louis. Mr. James Campbell, one of the principal promoters of

100 • THE GAS TRAMCAR

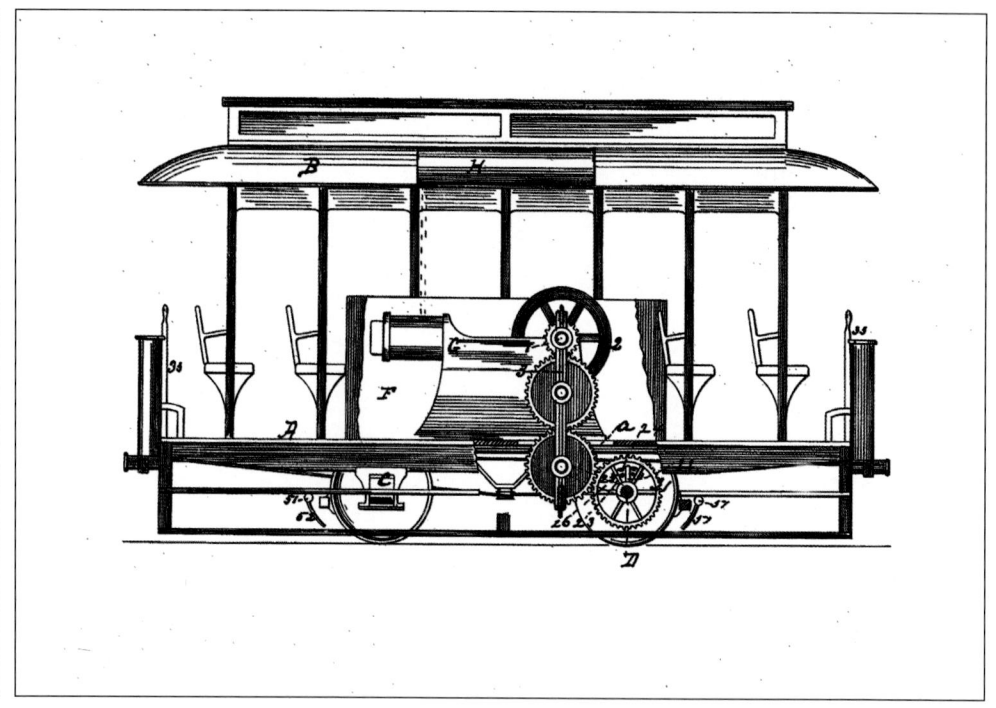

Right, below left and below right: The technical illustrations which accompanied Jay Noble's patent in September 1887. Despite the prototype having been built, tested, and reported in the press, no photographs of the car have been located. Turning the necessity of keeping the engine running all day into a benefit, Noble's patent included a small dynamo set, charging a wet cell battery while the car was stationary. This was used to generate electricity for the lights on the vehicle.

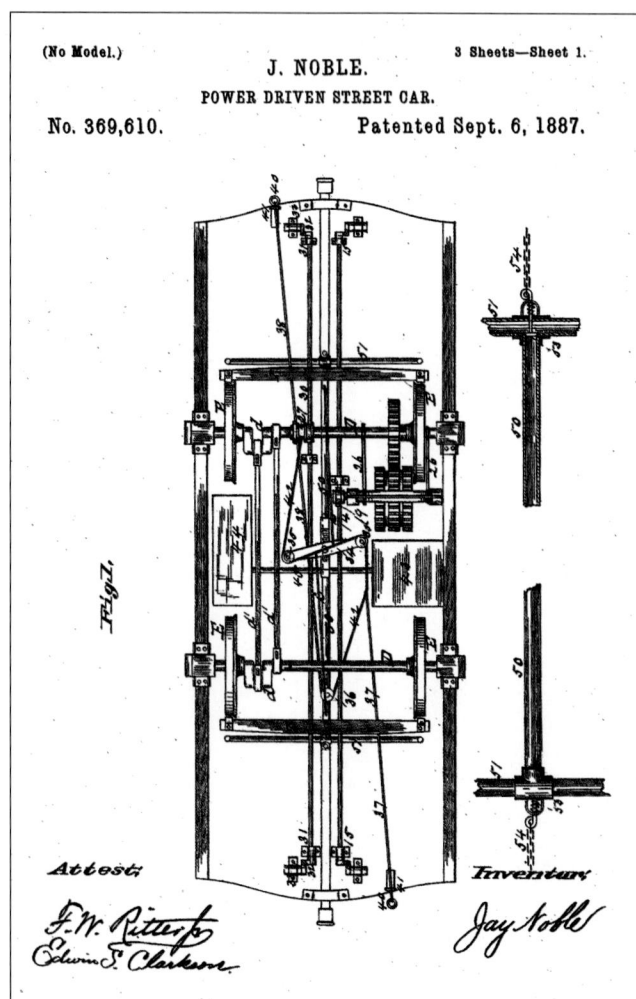

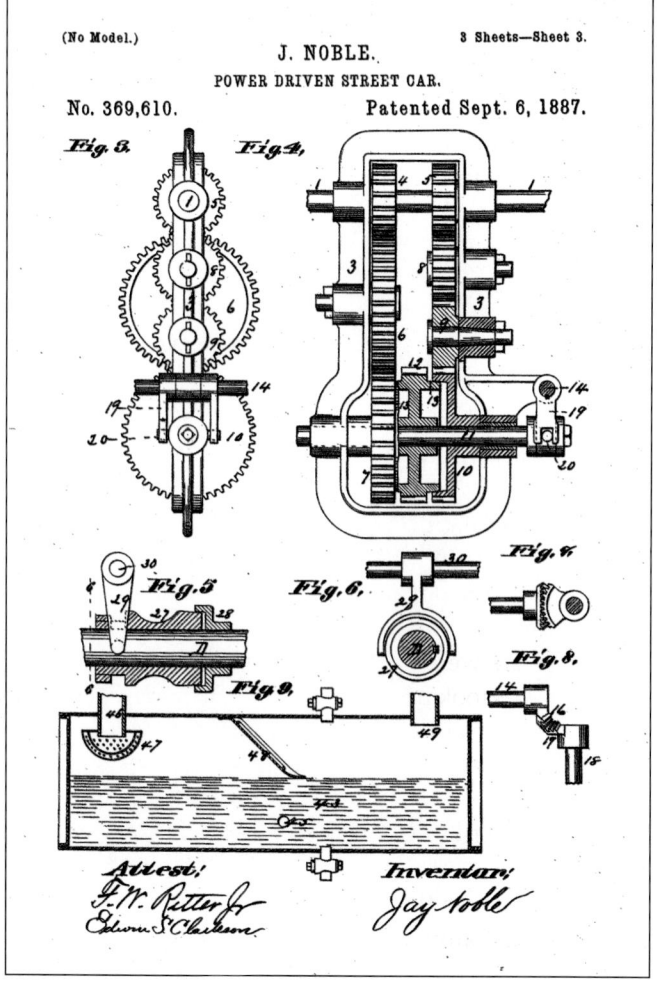

the Cross Town Road has the exclusive right to the use of the motor in Missouri. The new gas motor was first put on in Elizabeth in February 1889.

This motor was run over six months experimentally, developing abundant power for the heaviest loads and a speed of twelve miles per hour, but there were mechanical defects in construction, which were gradually eliminated, and two new motors containing marked improvements have been in use there during the past month.'

As with other designs for such a tramcar, the proposers realised that direct drive did not work with a gas engine, and incorporated a friction clutch and variable transmission, which they believed would enable them to use a much smaller and lighter-weight engine to power the car successfully without compromising performance. According to that 8 June 1890 newspaper report:

'The mechanism employed to accomplish this purpose is most ingenious. It is a friction device, exerting a powerful leverage, enabling an eight-horse power engine to easily start a loaded car on grades which could not be started by a thirty-horse power engine connected to the axle in the usual way. It is this device which makes this motor with a small light engine a success while others employing much larger engines proved failures.'

Although they were extensively tested, the Cross Town Electric Railway Company was never granted its franchise to build and operate the line, and no further records of the company have been found in the course of this project.

As neither of the proposed St. Louis gas tram systems ever entered into passenger-carrying service, that leaves Danks and Barnes in Melbourne as the first inventors to actually operate such a tramcar commercially. But were they the first with the idea? Not by several years.

Might that accolade go to Walter Andrew Harper and John William Rock of Oamaru on New Zealand's South Island who sought a patent for a gas tram in August 1882 three years before Danks and Barnes? They were encouraged by three New Zealand Acts of Parliament – the New Zealand Tramways Act of 1872, The Oamaru Town Hall and Gasworks Sites and Recreation Reserves Act of 1875 and the Oamaru Tramway Act of 1876.

The proposed tramway was to run from Severn Street in Oamaru to the Oamaru–Moeraki Railway, thus achieving two key aims – improving the town's access to the main rail network, and reducing the amount of traffic using the road between the town centre and the station. Although legally designated as a 'road tramway', it would in fact skirt the public park and run through waste land near the town centre which was scheduled for the development of a new gasworks.

There were as many objectors to a new gasworks being built so close to the town centre as there were objectors to the proposed tramway running through a park, and the two issues took up a lot of business time in the local council and in New Zealand's House of Representatives over a period of some years.

Deciding how the tramway should operate was another issue, and the one which, eventually, brought the Borough Engineer and the mechanical engineer John William Rock together. Their proposal to operate the tramway using gas-engined vehicles was a radical one.

This really was 'leading-edge' thinking on Rock's part, as no such vehicle had been operated anywhere in the world before that time. Gas engines were seen as offering a low-cost and clean future – promising greater efficiency than the most

To all to whom these Presents shall come We Walter Andrew Harper Surveyor and Engineer and John William Rock Mechanical Engineer both of Oamaru in the Provincial District of Otago and Colony of New Zealand ~~Engineers~~ send Greeting

Whereas we are desirous of obtaining Royal Letters Patent for securing unto us Her Majesty's special license that we our executors administrators and assigns and such others as we or they should at any time agree with and no others should and lawfully might from time to time and at all times during the term of Fourteen years (to be computed from the day on which this instrument shall be left at the office of the Patent Officer) make use exercise and vend within the Colony of New Zealand and its dependencies an invention of a new gas engine which may be used as a locomotive especially in propelling tram cars and in order to obtain the said Letters Patent we must by an instrument in writing under our hands and seals particularly describe and ascertain the nature of the said invention and in what manner the same is to be performed and must also enter into the covenant hereinafter contained **Now know ye** that the nature of the said invention and the manner in which the same is to be performed is particularly described and ascertained in and by the following statement (that is to say) The invention consists in the construction and use of a new gas engine which can be more easily and more economically worked than those now in use and which is capable of use as a locomotive more particularly for the purpose of driving tram cars. The gas used is to be stored in some place conveniently situated to the engine. In the case of a tram car it will be stored in a chamber under the seats as shewn in plan I. The engine resembles an ordinary steam engine except as regards the portions more particularly described. A general arrangement is shewn at Figure I. At Figure II is shewn a side view of the tramcar with the engine in the position which it would occupy when used with it Figure III shews a sectional plan of the cylinders and connections Figure IV is a transverse section through the cylinders while Figure V shews a sectional plan of the valves for admitting the gas and air to the cylinders. We will now describe the parts which are novel. A cast iron cylinder is represented at 1 Figure IV which is provided with a piston 2 of ordinary form fitted with packing rings 3, 4, so as to work air tight. This piston is moved backwards and forwards by means of an eccentric or crank upon the crankshaft of the engine and thus draws into the cylinder a mixture of gas and air through the valves 5 and 6, 6 is open to the air while 5 is connected with the reservoir containing the gas supply. The valve 5 is much smaller than the valve 6 so as to admit the due quantity of gas to air the proportion being about 1 to 10 or 12. Means for regulating this proportion are hereinafter described. The mixture of gas and air being drawn into the cylinder it is forced by the return stroke of the piston through the delivery valve 7 into the casing 8 of the cylinder 9 Here it remains under pressure until the slide valve 10 being moved by an eccentric on the crankshaft uncovers one of the ports 11 or 12. The compressed mixture immediately rushes through the opened port towards the cylinder 9 coming in contact on its way with the gas flame 13 Here it is immediately ignited and being very greatly increased in pressure a propelling force of great power is applied to the piston 14 which being connected in the usual manner to a crank shaft causes it to revolve and thereby imparts motion to whatever is desired At 15 and 16 are placed pieces of wire gauze protected on each side by perforated brass plates. These freely admit the gases while the ignition is prevented from passing into the casing 8. The gas burners 13 are connected with the gas reservoir and across each burner at a height sufficient for the flame to act upon it is placed a piece of platinum 17 This is heated to redness by the flame and reignites it if blown out When the expanded gases have forced the piston 14 to the end of the cylinder the slide valve 18 opens by similar means to the other valves and the waste gases are discharged through the obvious alternative – which was still steam traction of course – and much reduced infrastructure costs than that which would have been required had they chosen overhead electric pick-up, the latter system also requiring the construction of a power station.

As far as John Rock was concerned, having the Borough Engineer listed as 'co-inventor' on the Patent must have seemed to give the project a much greater chance of ultimate approval by the civic authorities in Oamaru.

At the time when the patent specification was submitted to the New Zealand Patent Office in Wellington, Walter Harper was Oamaru's Borough Engineer and both those considerations would have been at the forefront of his thinking.

Despite pre-dating Lührig's experiments with gas trams, aspects of Harper and Rock's specification are remarkably similar in many respects to those described in later patents in Germany, Australia and the United States – proof that when thinking logically, individual inventors living thousands of miles apart, and who are very unlikely ever to have been aware of each other's work, will frequently arrive at similar if not identical solutions to resolving the same challenges.

> 'The invention consists in the construction and use of a new gas engine which can be more easily and more economically worked than those now in use and which is capable of use as a locomotive or particularly for the purpose of driving tram cars. The gas used is to be stored in some place conveniently situated to the engine. In the case of a tram car it will be stored in a chamber under the seats.'

In the case of Carl Lührig, the decision to put the gas bags under the seats was arrived at only after earlier ideas to place them only beneath the floor had been tried and patented. In Harper's and Rock's thinking – and Danks' and Barnes' in Australia – under the seats was the logical solution from the outset.

The engine – or pair of engines to be precise as they were twin linked single-cylinder engines driving one pair of wheels – were to be slung below the chassis of the car, thus not obstructing the passenger compartment at all.

As Carl Lührig discovered several years later, trying to synchronise two engines would prove very difficult, and result in considerable vibration and a rather uncomfortable ride for the passengers.

The car was designed with a driving position at either end, but the specification does not give any detail as to how control might be switched from one end of the vehicle to the other.

Other aspects of their design were not fully articulated in the patent either. The absence of flywheels and clutches, for example, would seem to imply the engine was to be switched off by closing the gas supply valve every time the tramcar stopped, and restarted by re-opening that valve and re-igniting the gas as it passed over an open flame.

This would have been a procedure fraught with difficulties, resulting in a less than reliable passenger experience had the tramcar ever been built. Undaunted by such difficulties, the patent continued:

> 'The compressed mixture immediately rushes through the opened port towards the cylinder 9 coming in contact on its way with the gas flame 13. Here it is ignited and being very greatly increased in pressure a propelling force of great pressure is applied to the piston 14 which being connected in the usual manner to a crank shaft causes it to revolve and thereby imparts motion to whatever is desired. At 15

Opposite: The first page of the patent application for a gas tram submitted by Andrew Harper and John William Rock of Oamaru, New Zealand on 26 August 1882. This document is, at the time of writing, the earliest description found of a tramcar powered by a gas-motor engine. The application was scheduled for evaluation on 3 November 1882, but that hearing was postponed, and Harper disappeared in 1883 before the patent procedures could be completed.

> **Application for Patent.**
>
> Patent Office,
> Wellington, 28th August, 1882.
>
> PATENT for a new Gas Engine, which may be used as a Locomotive, especially in propelling Tramcars.
>
> WALTER ANDREW HARPER, Surveyor and Engineer, and JOHN WILLIAM ROCK, Mechanical Engineer, both of Oamaru, New Zealand, have deposited at this office a specification of the said invention; and I have appointed Friday, the 3rd day of November next, at 10 o'clock in the forenoon, at this office, to hear the said application and all objections thereto; and I require all persons having an interest in opposing the grant of such Letters Patent to leave, on or before the 30th day of October next, at this office, particulars in writing of their objections to the said application, otherwise they will be precluded from urging the same.
>
> W. S. REID,
> No. 681.   Patent Officer.

Harper and Rock's application for a patent for their gas tramcar was scheduled for evaluation on 3 November 1882. The hearing was advertised in the *New Zealand Gazette* but was postponed, and Harper disappeared in 1883 – it remains unclear whether or not the patent process was actually formally completed.

and 16 are placed piece of wire gauze protected on each side by perforated brass plates. These freely admit the gases while the ignition is prevented from passing into the casing 8. The gas burners 13 are connected with the gas reservoir and access each burner at a height sufficient for the flame to act upon it is placed a piece of platinum17. This is heated to redness by the flame and reignites it if blown out.'

The specification of the vehicle in the patent application was clearly incomplete, with several key issues still to be full articulated, suggesting strongly that Harper and Rock had only a rudimentary understanding of the requirements of working a gas-engined vehicle. For example, there is no mechanism identified for stopping or starting the vehicle without shutting the engine down and then restarting it.

The patent application was suspended in late 1882 and in 1883 Harper took three months leave of absence from his post as Oamaru's Borough Engineer and subsequently disappeared.

Development work on the tramcar and the proposed tramway – a project he had invested so much time and effort into – ceased, and it had faded into obscurity until rediscovered during the research for this book.

In Europe in the 1890s, however. it must have felt as if the gas tram could look forward to a very bright future – yes there were still major power issues which needed to be resolved, but they would surely not turn out to be insurmountable – gas engines were getting more powerful and more compact.

In his conclusion to his lecture to the Société des Ingenieurs Civil de France in 1896, Monsieur Lavezzari described the currently available vehicles in detail and, after enumerating the anticipated costs of operating such a tramway, he offered his audience a summary of what was, at the time, the current stage of development of gas traction.

At the same time he put forward what he believed was a possible roadmap for such a system's future development:

'The consumption figures that I have just indicated demonstrate that gas traction deserves an honorable place among other systems. It is a mode of operation which is easy to establish almost everywhere, inexpensive and safe. The central factories and the personnel are not very important.

Almost all the inhabited centres have gas works, so this point cannot constitute a serious difficulty; it could hardly come from the selling price of gas in certain towns; but it will certainly happen that the Companies will make special reduced rates. However, if in certain cases they did not want to or could not, the Company

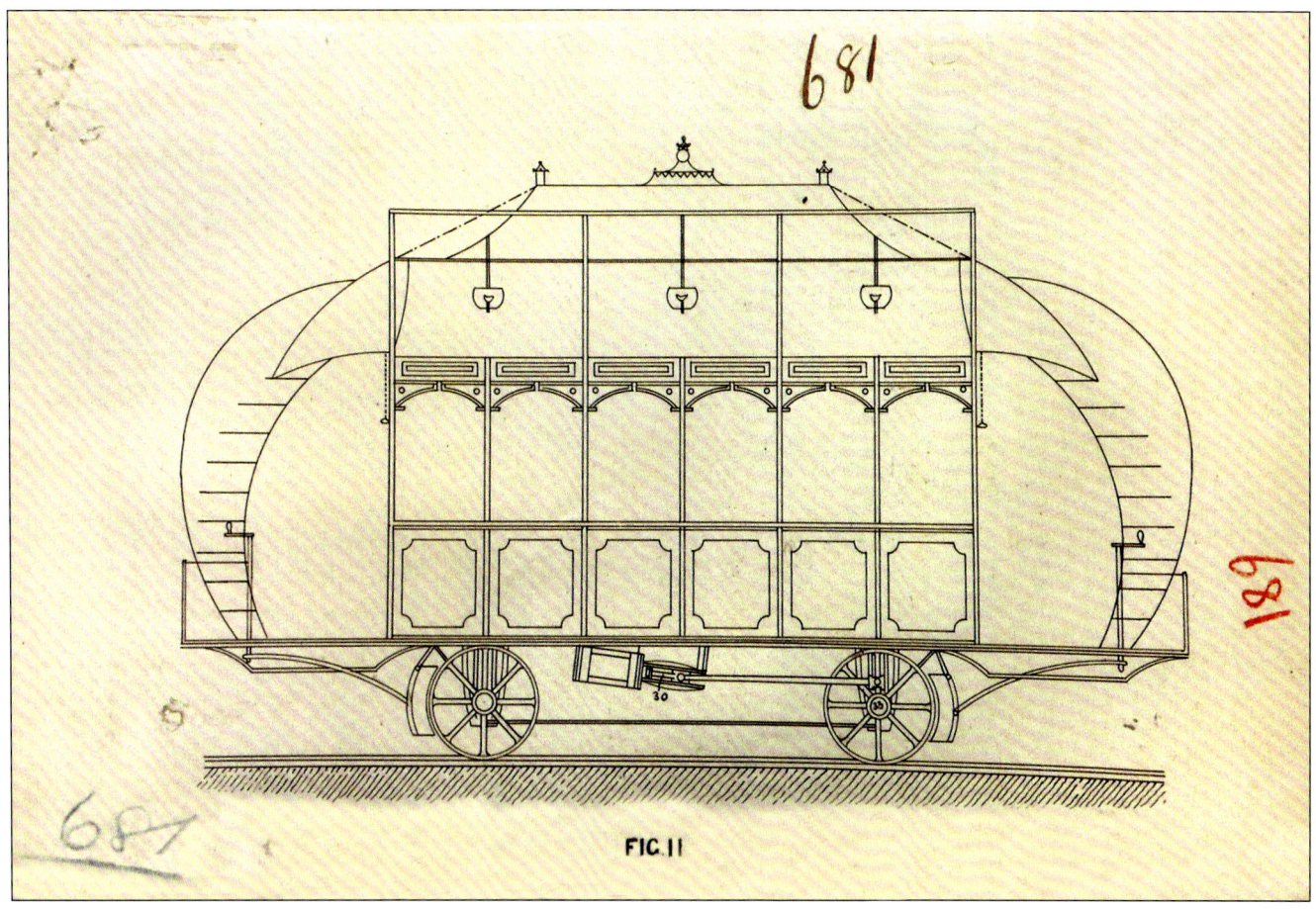

FIG.11

which would operate the tramway would have the resource to manufacture its own gas, then low grade gas would be acceptable. It would be perfectly suitable for this use, and moreover its manufacture is simple and inexpensive. But this system, still young, has to make progress like its predecessors, in order to reach its last degree of perfection.

For my part, I only see the current cars as a useful demonstration that the problem is worth investigating, but I believe that to make it a truly practical car there is still a lot to be perfected.

Currently we have been content to take an ordinary engine that we have fitted almost without modification on the cars; it is almost like building a locomotive with a stationary steam engine; the needs are not the same, and different machines are obviously needed.

The main drawback of these engines is that they cannot cope with steep slopes; or the alternative is to fit engines which are too powerful to control in ordinary circumstances. It would also be good to be able to completely stop the engine when the car stops, to avoid the vibration and unpleasant noise that one currently experiences.

I believe that we must look for improvement, and perhaps one day, if progress meets my expectations, I will have the opportunity to talk to the Society about it.

Finally in the mechanical part of the transmission, there are also some improvements to be made; the gears make a rather unpleasant noise during the whole walking time; we could perhaps remedy it immediately very simply by using wooden-toothed gears.

Harper and Rock's tramcar, as illustrated in their application for New Zealand Patent No.681. While the patent specification makes no mention of a double-decked vehicle, that is clearly what they were proposing. It is unlikely that two such small single-cylinder gas engines could have powered a fully laden car with this sort of capacity. As they were not linked, balancing their output would have been very difficult and likely to cause severe vibration – as would later be found with Carl Lührig's twin-engined car.

Right: 'Fig.I' from the drawings supplied with Harper and Rock's patent application. The specification describes the working of the engine thus – 'This piston is moved backwards and forwards by means of an eccentric or crank upon the crankshaft of the engine and thus draws into the cylinder a mixture of gas and air through the valves 5 and 6. 6 is open to the air while 5 is connected with the reservoir containing the gas supply. The valve 5 is much smaller than the valve 6 so as to admit the due quantity of gas to air the proportion being about 1 to 10 or 12.'

Below: Figures III, IV and V from Harper and Rock's patent application show parts of the valve assembly.

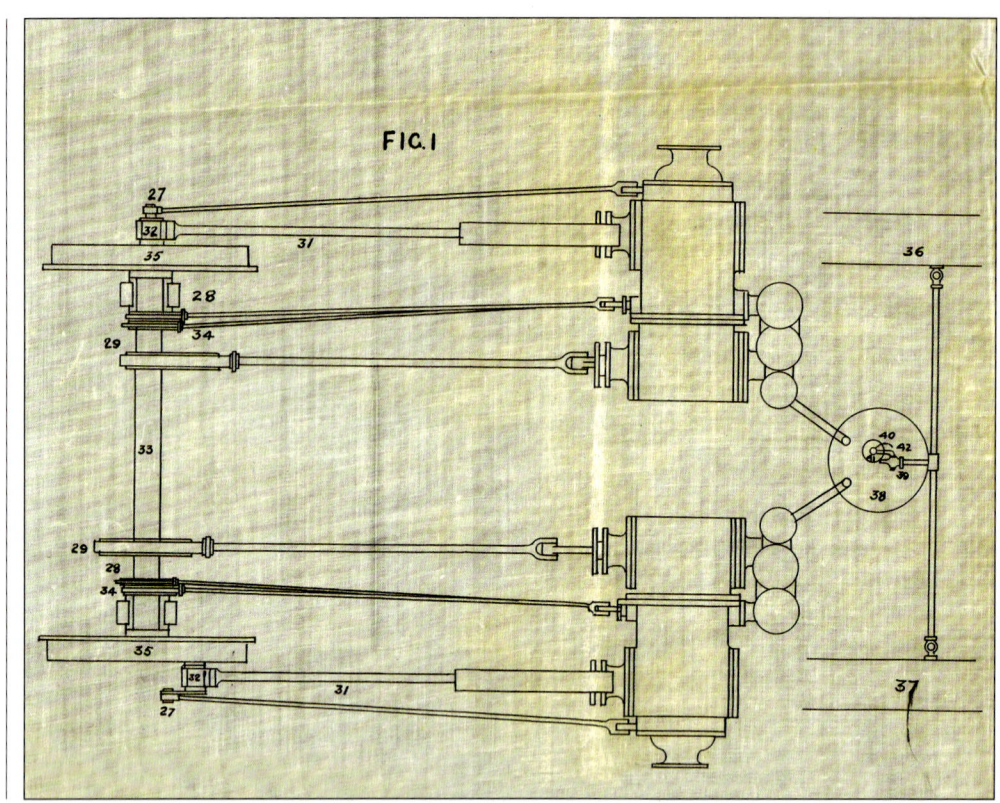

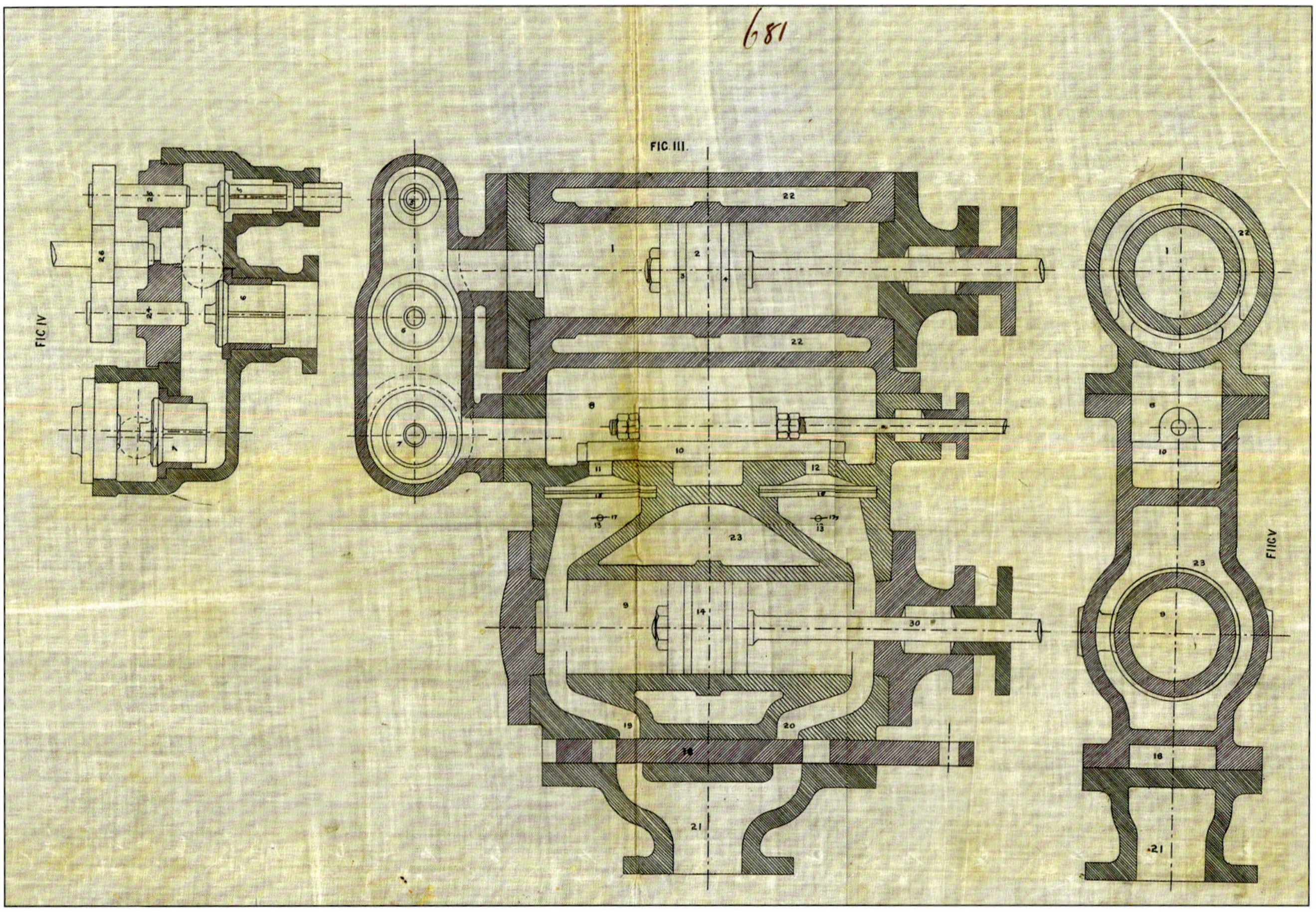

This is just an overview of the first improvements to be made; the field is open to inventors, and we will be happy to record the progress that will be made and which will make it a superior model and, we must hope, of French origin.'

Lavezzari would doubtless have been disappointed that Paris never took up the challenge of gas trams, opting instead for a mixture of steam, cable and electric – especially as one of the early enthusiasts for gas tramways was fellow Frenchman Jean Marie Armand Montclar.

At the time of the granting of his French Patent on 10 December 1881 – No.146,290 *Gas-Locomotor for the Locomotion of Vehicles, Carriages, Tramways, Wagon &c.* – Montclar described himself as a 'civil engineer of Paris', but fifteen years earlier, in the 27 March 1866 issue of *The London Gazette*, when seeking patent protection for '*Improvements in the Manufacture of Materials or Compositions for Decolorizing or Purifying Saccharine or other liquids, and for making paint, blacking, and foundry blackening, and in apparatus therefor*', he had described himself as an 'engineer, of Java, but at present residing in Glasgow, in the county of Lanark, North Britain'.

At the heart of his tramcar was to be an Otto-cycle engine with one or more cylinders, and he anticipated the system being used either as a separate locomotive to haul former horse cars, or with the engine laid horizontally beneath the tramcar floor. He was granted several patents for the vehicle and its engine – in addition to the French Patent, it was patented in Belgium on 15 December 1881 (Patent No.56,500), in Britain two days later (Patent No.5,534) and in the United States on 13 June 1882 (Patent No.259,413). Unusually, Montclar seems to have intended for the gas supply to be cut off whenever the tram was required to halt, but he expressed great faith in the effectiveness of his system, and unusually for a patent specification, introducing some anecdotal remarks:

'The gas-motor engines of my invention resolve in the most satisfactory way the important question, namely, which is the most convenient way of locomotion in towns? They are noiseless, and provided with an invisible escape, so there is no risk of frightening horses by whistling or other unusual noise.'

No records of the tramcar having been built and tested have so far been located, but in their day Montclar's plans were widely enough reported for him to have been listed as one of the gas tram 'pioneers' in *De Gasmotortram volgens het Systeem Lührig*, published in 1896 by 'The Gas Traction Company departm, Netherlands and its Colonies' based in Amsterdam, who were actively promoting Lührig-type gas tramcars for Amsterdam..

A rather more eccentric patent was granted to James Morris O'Kelly of New York ten years later in 1892. He gave only a very sketchy description of the engine he planned to install in his vehicle, which would drive the front pair of wheels, but a lot more detail about the reversible seating – an unusual addition on a vehicle with the engine and driving position at one end only – and the provision of removable parasols for passengers on the upper deck!

Five years before O'Kelly, William C. Carrick of Philadelphia had patented his own idea of a gas-engined tramway but, like Montclar he elected to design a vehicle where the engine was in a separate 'dummy car' hauling a tramcar. The unique nature of his proposal was that the gas supply was to be carried in tanks underneath the tramcar's seats – as would become the norm in Lührig's cars – but fed to the engine by a flexible coupling.

108 • THE GAS TRAMCAR

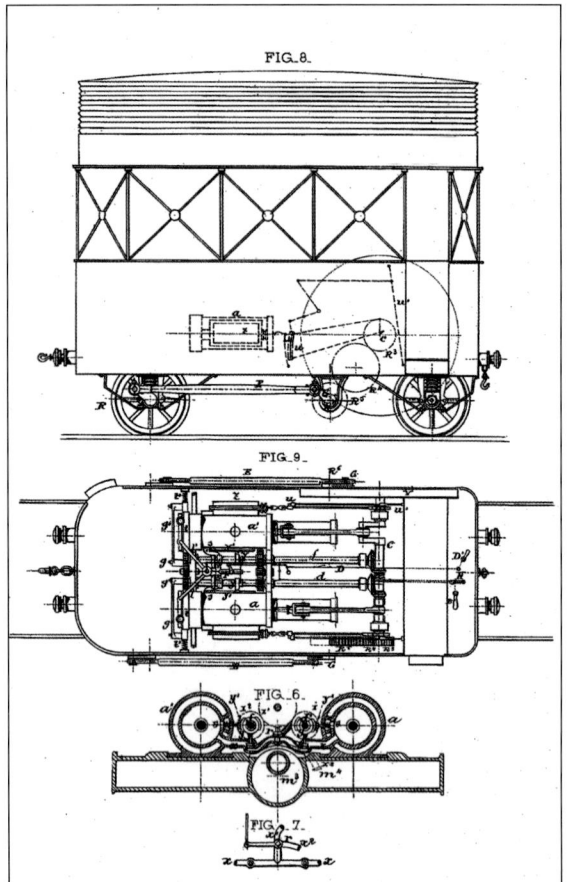

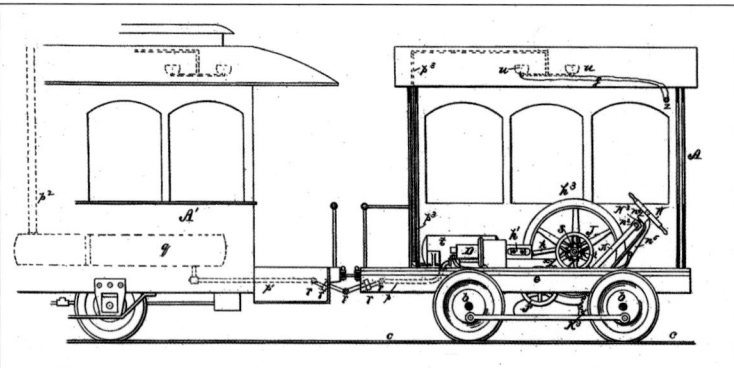

Danks and Barnes seems to have been the first to carry fare-paying passengers but even theirs was not the first such proposal – preceded on paper at least by the plans of Conrad Krauss in 1878, Joseph and Joseph Quick in 1881, and Walter Harper and John Rock the following year.

There were certainly several other inventors whose ideas never got beyond the drawing board and have yet to be rediscovered.

With the luxury of hindsight, what is perhaps surprising is the length of time which elapsed between the first proposed application of a gas engine to a tramcar, and the opening of the first tramway to operate gas tramcars commercially. Developments in Victorian engineering usually progressed at a much more accelerated pace.

But, as has been seen in this chapter, there were persistent concerns about the power – or lack of it – in those early gas engines which were small enough to be fitted in tramcars. While the theory of employing gas traction on tramways was sound enough, the emerging technology was slow to catch up.

Despite the fact that the first commercial gas-fuelled tramways operated in Australia and Germany, we must look back to Britain in the late 1870s in order to explore the lineage and development of Lührig's gas trams.

Above: Jean Marie Armand Monclar's initial design, patented in 1881, was for a two-cylinder gas-engined locomotive intended to haul former horse tramcars.

Above right: William Carrick's 1887 proposals envisaged the gas being stored under the tramcar seats – in the manner later adopted by Carl Lührig – but then being fed to the separate locomotive through a flexible hosepipe.

Right: A detailed drawing of Carrick's gas engine locomotive. Like Lührig, he envisaged a horizontal engine which would have given the vehicle a low centre of gravity and greater stability.

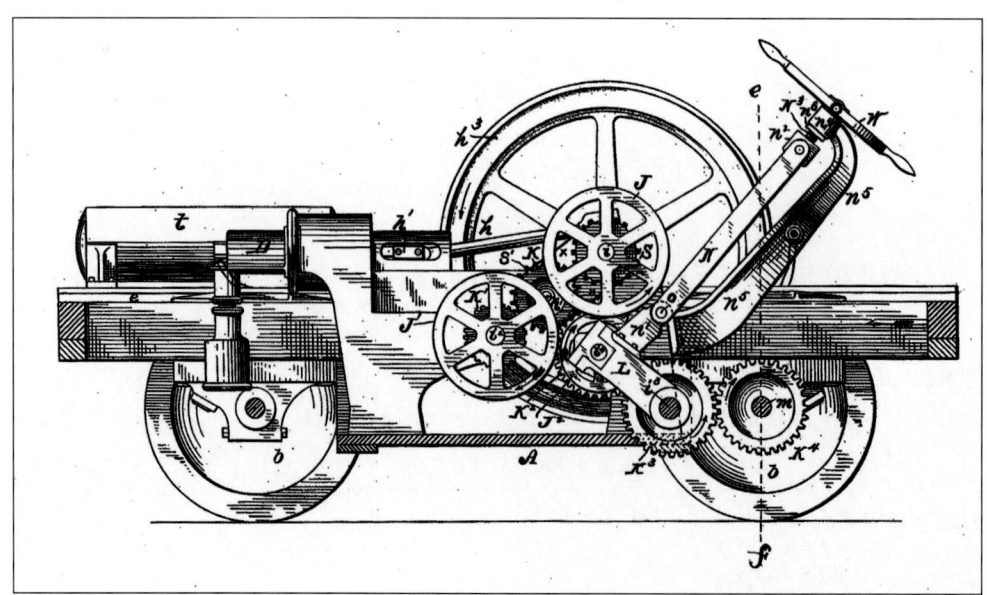

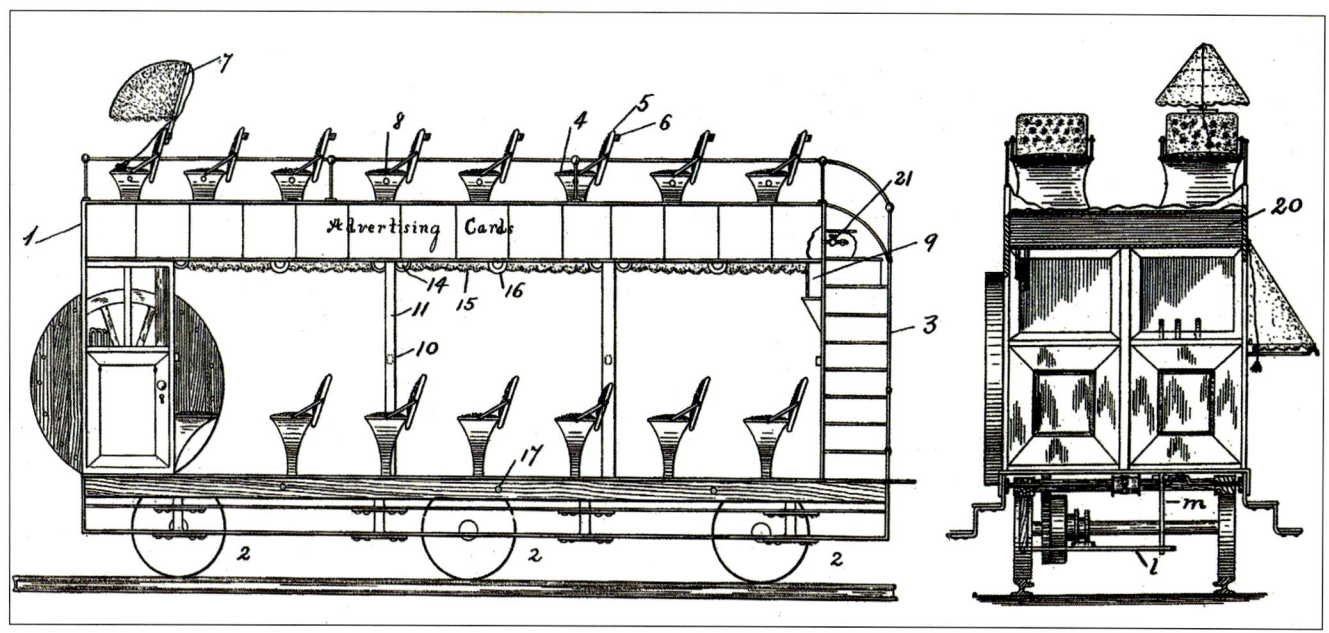

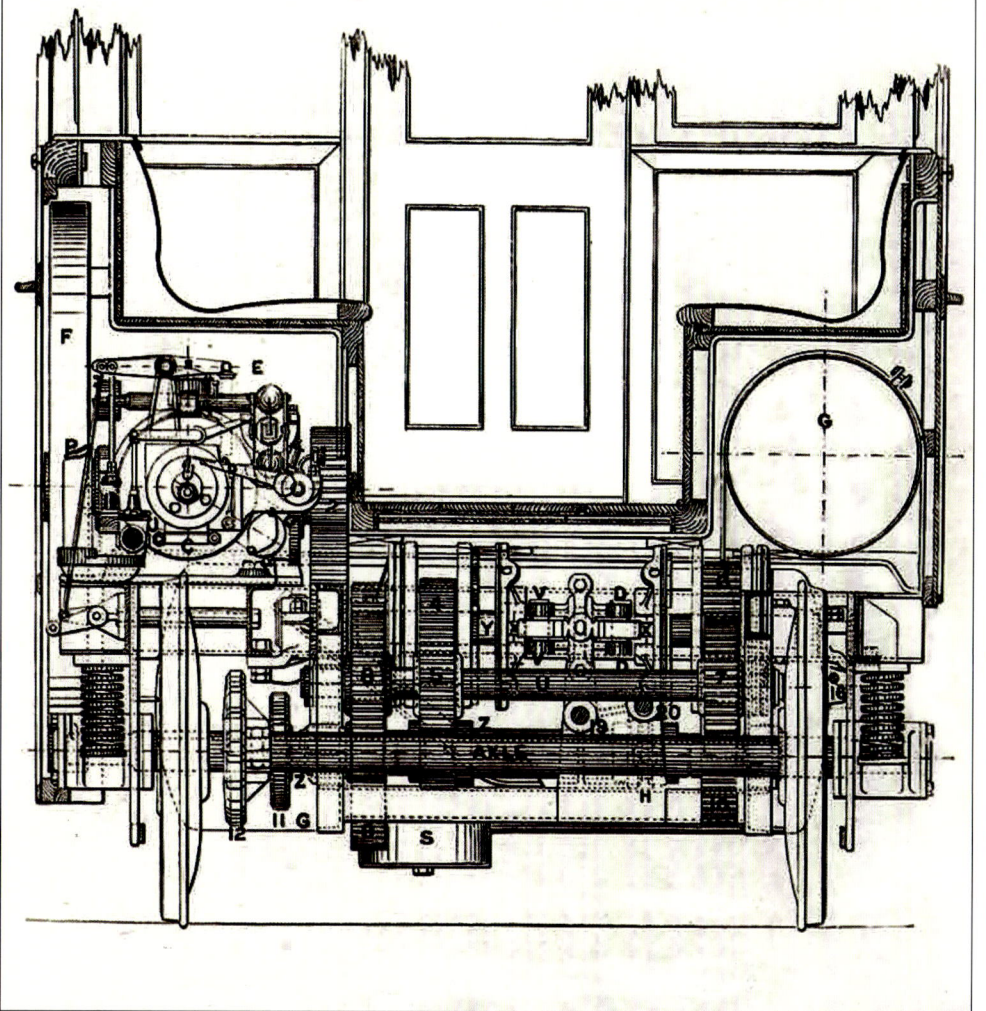

Above: The drawings which accompanied James O'Kelly's 1892 patent for a tramcar were highly fanciful, and clearly had not been fully thought through.

Left: This very detailed end-view of the complex gearing which would be used in the British Gas Traction Company's tramcars introduced on the Blackpool St. Anne's & Lytham Tramway is in sharp contrast to the very sketchy details contained in several contemporary patents from other tram designers. This illustration appeared in *The Engineer* on 28 July 1899. The engine and gearing viewed from the opposite end of the vehicle is illustrated on page 9.

Holt et Crossley. Fig. 1 à 10.

Fig. 7.
Fig. 9.
Fig. 8.
Fig. 1.
Fig. 4.
Fig. 10.
Fig. 2.
Fig. 3.

Holt et Crossley.

Fig. 11.

THE GAS TRAMCAR IN BRITAIN

Throughout the relatively short history of the development of the gas-engined tramcar, a small number of names dominate. Those of Nicolaus Otto and Carl Lührig in Germany were key to bringing the gas tramcar into useful service, but as has been seen, the original ideas predate Lührig's involvement by several years, and can be traced back to just two years after Otto first demonstrated his four-stroke engine.

At the forefront in Britain were the Crossley Brothers in Manchester, Francis William (1839–97) and William John (1844–1911); and Henry Percy Holt (1848–1902). Holt's involvement with gas trams – and the companies operating them – endured until his death. Both Crossley Brothers had served engineering apprenticeships, and Francis – known as Frank – had worked at Robert Stephenson & Company at Elswick in Newcastle before going into business first with John Macmillan Dunlop and later with his brother William. William was the businessman, Frank the engineer.

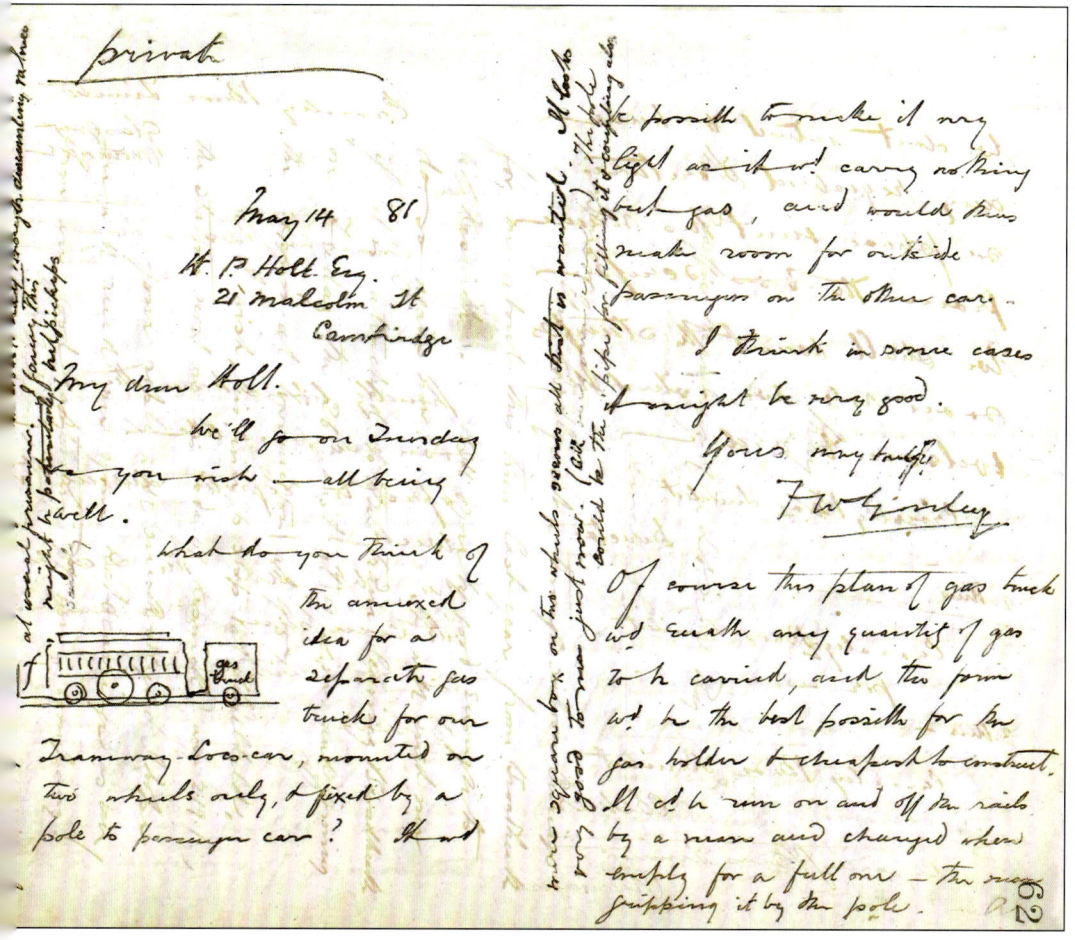

Opposite: Drawings from 1878 to 1881 of the engines and transmission of Holt and Crossley's gas tramcar and a gas locomotive for hauling former horse cars illustrated in *Les Moteurs à Gaz* by Gustave Richard, 1885.

Left: F. W. Crossley's letter to Henry Holt outlining his idea for a separate 'gas truck' for his tramcar. By this stage the two men had several ideas for the mechanics of the tram.

The Crossley Brothers' engineering company was a relatively small enterprise until, in 1869, they acquired the British and Colonial rights to the Otto/Langen engine direct from Otto himself. They later also acquired the British and Colonial rights to Otto's four-stroke engine, but as the development of that engine progressed, and its application to tram traction evolved, it would appear that their licence was not an exclusive one, despite them later jointly developing and patenting improvements to the engine with Otto's Gasmotoren-Fabrik Deutz.

Frank was an innovative engineer and his improvements to Otto's original configuration of the engine greatly improved its efficiency and output. He was also keen to identify and develop commercial applications for the engine.

Crossleys started building Otto-cycle engines in 1876 and by the following year these engines, in an increasing range of sizes and output, very quickly formed the majority of their output.

For most people involved in small to medium sized industries, the gas engine offered a more cost-effective alternative to the steam engine which had dominated the manufacturing sector for more than half a century, but Frank Crossley could clearly see a future for the device as a replacement for both horses and steam power on tramways.

He and Henry Holt appear to have started discussing the potential of developing a gas-engined vehicle as early as 1878, and filed a Provisional Patent Specification No.1912 in May 1879 titled *Improvements in Machinery for Starting, Propelling, and Stopping vehicles, and in the apparatus and appliances connected therewith, more particularly with reference to gas engines &c. &c.*

They did not, however, proceed to submitting a full patent application, instead submitting a revised and more complete proposal in November 1879 which became GB Patent No.4499.

Nineteenth century patents were often granted without the sort of meticulous searching you would expect, and were subsequently often the cause of litigation to establish who held the primacy of ideas. But in such a rapidly evolving arena, ideas were often patented and quickly usurped by later developments – often by the same people.

The side elevation and transverse section illustrations for the gas-engined locomotive unit described in Holt and Crossley's Patent No.4499, 4 November 1879. Hydraulics were used to raise and lower the engine-carrying frame and driving wheels to control motion. To increase speed, hydraulic pressure controlled how closely in contact the driving wheels were with the rails. They claimed patent rights for 'starting, propelling, stopping and reversing the direction of motion of locomotive vehicles, of the machinery, apparatus, and appliances, substantially as herein described, whereby the driving axle can be lowered or raised, and turned to an inverted position relatively to the vehicle, so as to vary the grip of the driving wheels, and to stop or reverse the motion of the vehicle without stopping or reversing that of the engine which drives it.'

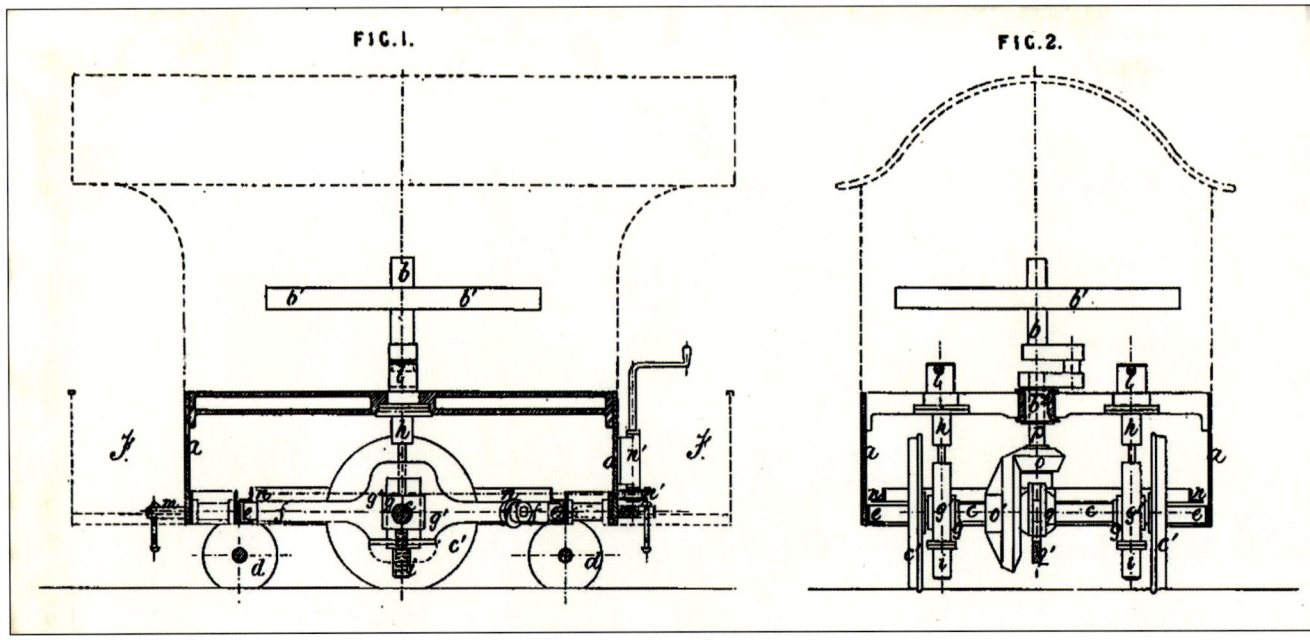

Early on, Crossley and Holt were alert to the options available to them. The patent included a design for a locomotive which could be used to haul former horse trams or an integrated vehicle, and also contains several innovations which would later also be covered by patents from Carl Lührig and others:

'In this invention the motor engine may be contained within or on the passenger vehicle itself, or arranged as a separate locomotive…
  We also generally place the driving wheels between two pairs of carrying wheels, but their position is capable of being changed as circumstances may require.'

The engine was unusual in that the entire engine assembly – including the large driving wheels which were to be kept running – was to be raised and lowered as required using hydraulics, making contact with the rails when moving, but sitting slightly above the rails when stationery.

Equally unusual was the idea of placing the engine on its side in one of their tramway locomotive designs, the flywheel rotating horizontally rather than vertically. The drive shaft from the flywheel ran down to the 'driving axle' where it meshed with gears on the axle.

A letter from Crossley to Holt survives in the Crossley Archives dated 14 May 1881, just a few weeks before Holt joined the staff of Crossley Brothers, and is clearly just one in a sequence of exchanges between the two men as they evolved their ideas for a gas-engined tramcar.

In the letter, that basic configuration mirrors the patent – two large wheels connected to the engine by a system of gears and clutches, with the 'carrying wheels' front and back. We can hypothesise that this shared interest in exploiting the potential of gas traction was part of the reason why Holt subsequently joined the company. The letter includes one of the earliest illustrations of their idea how a passenger tramcar with an integral engine might eventually look like:

'May 14 1881
My dear Holt,
We'll go on Sunday as you wish – all being well.
  What do you think of the annexed idea for a separate gas truck for our tramway loco-car – mounted on two wheels only, + fixed by a pole to the passenger car?
  It wd be possible to make it very light as it would carry nothing but gas, and would thus make room for outside passengers on the other car.
  I think in some cases it might be very good.
                                        Yours very truly F. W. Crossley

Of course this plan of the gas truck wd enable any quantity of gas to be carried, and this form wd be the best possible for the gas holder + cheapest to construct.
  It cd be run on and off the rails by a man and changed when empty for a full one, the man gripping it by the pole.'

Making room for 'outside passengers' might imply that the two men were possibly thinking about a double-decked vehicle. In the margin notes, Crossley also looks forward to the possibility of using compressed gas, which had already been proposed – and patented – by Krauss and the Quicks, and which would become the norm in all the operational gas-engined trams. Had his idea of a gas trailer containing compressed gas been built, the range of the proposed tramcar could have been considerably

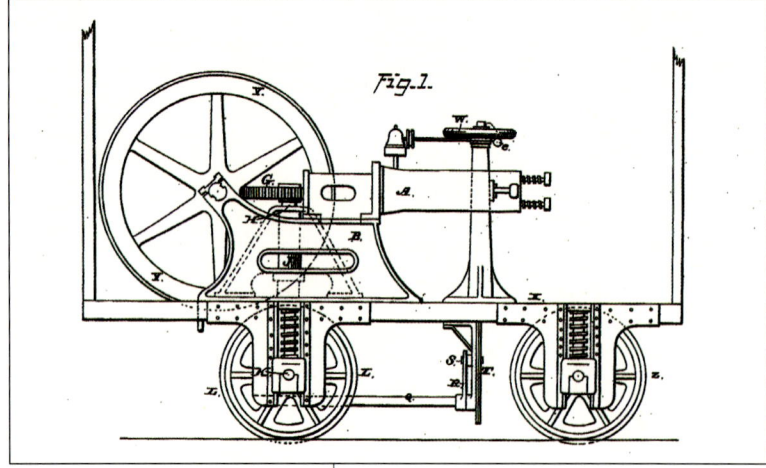
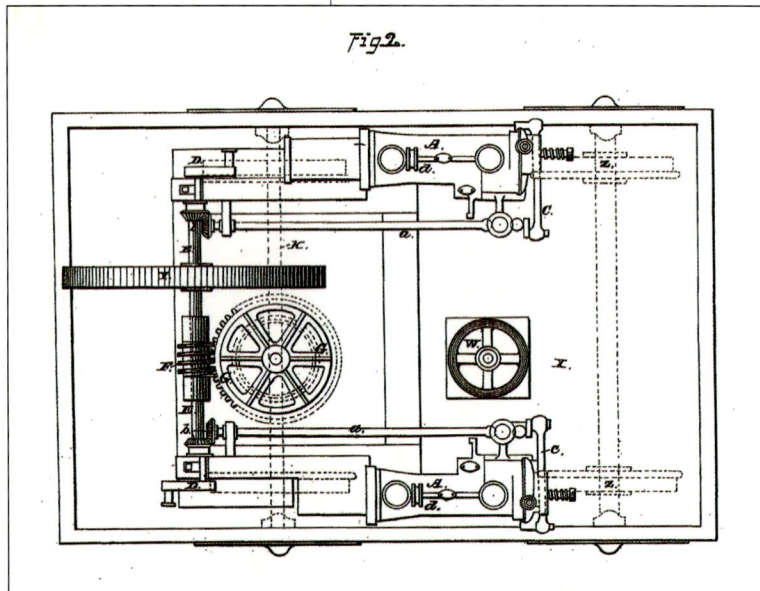
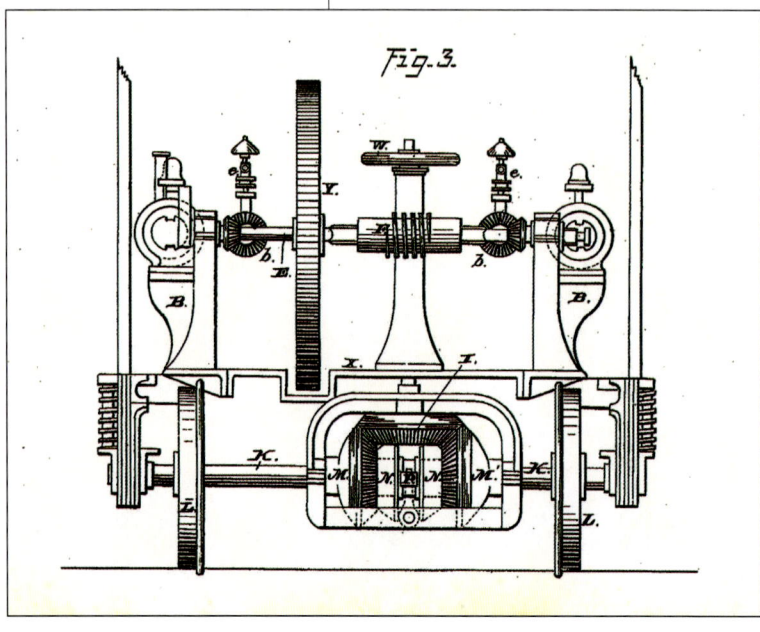

extended – but the vehicles would have been more complicated to turn at the end of the journey.

However, despite their enthusiasm, Crossleys do not seem to have progressed the idea, although Holt's interest remained undiminished, leading to his involvement with the Gas Traction Company a decade later.

Another gas-engined tram locomotive was patented on 10 August 1879 by John Roger Purssell of Blackfriars, London, who envisaged two single-cylinder water-cooled engines, their cranks driving a single shaft on to which was mounted a large flywheel and a worm drive, the latter being engaged or disengaged from a horizontal toothed wheel as required – thus converting the rotation of the drive shaft, via a vertical shaft, to a horizontal bevelled gear-wheel and clutch mechanism driving one of the axles.

He was proposing a much more complex drive train than many of his contemporaries, but the actual vehicle does not appear to have ever been built.

This seems to have been very much a 'work in progress' at the time the patent specification was filed as there was no provision in any of the accompanying diagrams for how the gas supply was to be stored or delivered to the engines – although the patent specification did note that it would be kept in pressurised tanks beneath the floor.

Those issues had still not been fully articulated when he took out an American Patent – US Patent No.231,097 – on 10 August 1880, but he did, however, note that he intended to use:

'… "silent gas-engines" …supplied with gas, which can be kept in a compressed form within strong tanks fitted under or upon the platform in any convenient manner, the water-tanks being preferably over the rear wheels, where dead weight is necessary, as will be well understood.'

On 21 October 1893, the first run in Britain of a small twelve-seater Lührig gas-engined tramcar took place, but the inspecting officer Major-General Charles Scrope Hutchinson,

Chief Inspecting Officer for Railways at the time, expressed severe misgivings about its design and safety. Significant concerns related to the inefficiency of the brakes, and the fact that there was no indication of which way to turn the wheel to apply the brakes.

Hutchinson required direction arrows to be clearly painted on the wheel. There was also a foot pedal on each platform which released and engaged the clutch, and on the test car, there was no way to lock that pedal on whichever was the rear platform at the time depending on the direction of travel.

General Hutchinson insisted that locks be installed to ensure passengers did not accidentally engage or release the pedal. He also required 'life protectors' – sometimes known as 'cow catchers' – to be fitted at either end of the car, and these subsequently became a standard feature of all British gas trams.

The next trial of a gas tram in Britain took place, also in Croydon in 1894, and appeared in an article in *The Engineer* in its issue for 22 June 1894, headed 'A GAS MOTOR TRAMCAR'.

> 'On Tuesday last the Traction Syndicate Limited ran their new gas-motor tramcar for a short trip on the Croydon and Thornton Heath Tramway Company's system. As this mode of traction is somewhat novel, a brief description of this system may be of interest. The invention hails from Dresden, the designs of Herr Lührig having been greatly improved by Mr. G. P. Holt, of Crossley Bros., to whom is due the credit of bringing the car to a state of perfection which bids fair to make a commercial success.'

'G. P. Holt' was, of course, a misprint, the gentleman in question being Henry Percy Holt who, thirteen years after his initial exchange of ideas with Frank Crossley, had finally seen a gas tram successfully put through its paces.

There is a confusion of dates here, for the specification of the 'new gas-motor tramcar' being referred to – 18 feet long and accommodating twenty-eight passengers, substantially larger than the original Dessau cars which had a capacity of twenty, fourteen seated, six standing – had actually been given its first trial run on 1 May.

It was a single-decked car of the size used in Dresden, fitted with a Deutz engine which had been modified according to Holt's specification. The article noted that distinctive features of the car were its cow catchers or 'life-preservers'.

But Croydon's interest in gas traction had already been piqued with some test runs in early 1893 of an American-designed Connelly gas motor locomotive, intended to pull former horse trams.

John Connelly, from Plainfield in New Jersey, had developed and patented his first oil- or gas-engined tramcars in March 1886, fitted with an oil engine and large flywheel at one end, and a succession of patents followed.

Two further American patents, also from 1886 – Nos.345,279 dated 13 July and 347,470 dated 17 August – both described street-cars with integral gas motors, and in the first of them, he explained the rationale for his invention:

> 'The expense attending the maintenance of horse street-car lines has led to many attempts to substitute in place of horse-power some suitable motor to be carried by and to drive the cars; but none of such devices have come into general use. Among the reasons for this is, that the power required to start a heavily-laden street-car or to drive it up a grade has been thought to necessitate the use of a powerful engine,

Opposite: Three illustrations of Purssell's 1879 patent gas tram car, from Gustave Richard's *Les Moteurs à Gaz* published in 1885. In his patent, he is clearly anticipating a much more 'modern' drive train than several of his contemporaries.

116 • THE GAS TRAMCAR

Right: The Connelly Motor as illustrated in *The Street Railway Journal* in 1886, showing the roof-mounted gas storage tanks.

Below right: Connelly's proposed belt and chain drive, as illustrated in his 13 July 1886 US Patent No.345,279.

Below: In his 17 August 1886 US Patent, No.347,470, Connelly used a more sophisticated design with a small engine driving just one pair of wheels, but with four selectable gear ratios to be used when starting from stationary.

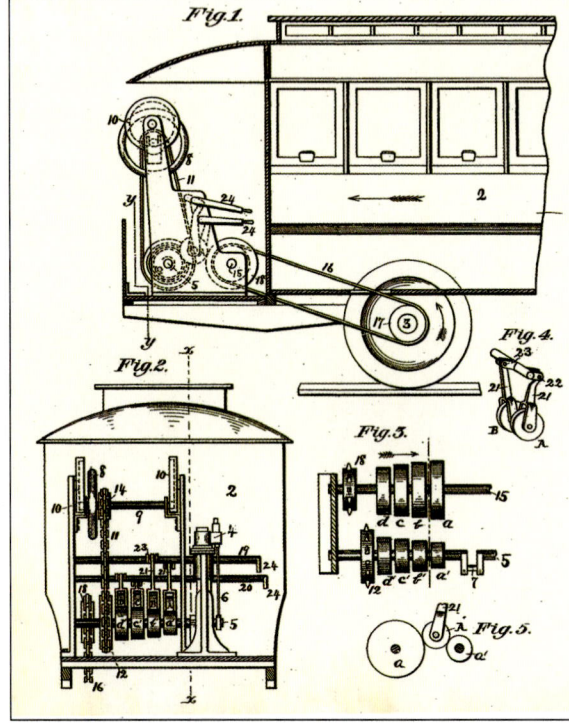

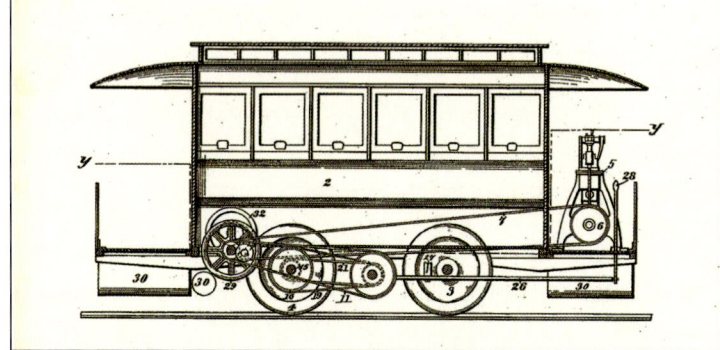

much more powerful than necessary to drive the car after it has been started on a level track; hence the engines heretofore used for the work have been too heavy, and have entailed much expense to render their adoption desirable.'

He expanded his thinking in the August 1886 patent, while at the same time introducing much more sophisticated gearing (see image) into the intended vehicle, to further help overcome inertia when moving away from stationary:

'The object of the invention is to provide a motor apparatus for use on street-cars of such construction that engines of small power may be used successfully, and with a saving of expense as compared with the present common use of horse-power and cable lines. As in the device for which I have already made application for a patent, I use on each car a continuously-running engine, together with devices for throwing the engine in and out of gear with the driving axle of the car, and thus starting or stopping it. The continuous motion of the engine while the car is stationary stores sufficient energy to overcome the inertia of the car and enable it to be started with ease.'

A very rare photograph of one of the Connelly gas-engined locomotives during its trials in England. This vehicle is believed to have been built under licence by Weyman & Company of Guildford and was run for six months in 1892–3 on the Rotherhithe Road section of the Deptford and Greenwich Tramway – the first tramway in Britain to carry fare-paying passengers on a tram hauled by a gas-engined vehicle. It was also briefly tested on the Croydon and Thornton Heath Tramway. The gas-engined locomotive had been patented by John S. Connelly as US Patent No.426,985 on 29 April 1890, describing a vertical single-cylinder engine, whereas the Weyman variant had twin vertical cylinders. Connelly had already patented several tramcars in 1886 and 1887 with integrated engines – see opposite – which could be fuelled with oil or gas, but this was the first which he specifically patented as a Gas Engine.

In an appendix to the *Proceedings of the American Gas Institute* in April 1894, the patent gears were highly praised:

> 'A very large part of its satisfactory working is due to an exceedingly ingenious and simple mechanism for transmitting the power to the axle, whereby, without altering the speed of the engine the power may be slowly applied for starting or

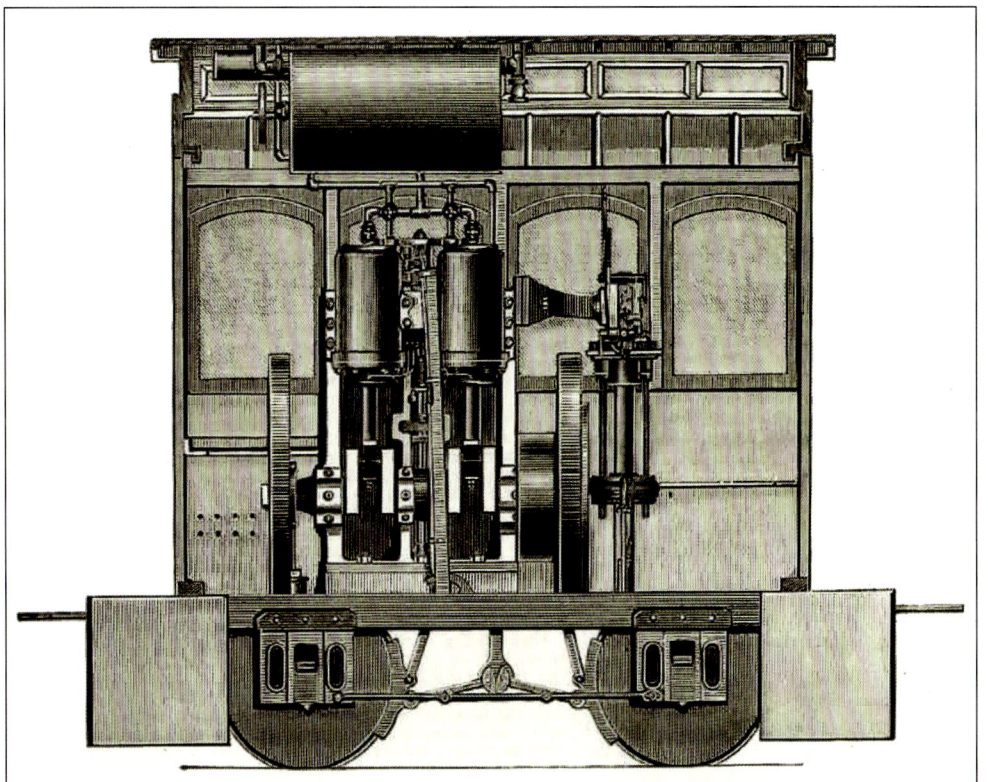

The Connelly Gas Tramway Motor of 1891 was featured in the issue of *The Engineer* for 30 September 1892 to coincide with the introduction of an improvement on it which could run on a range of gases including vapourised gasoline and vapourised naphtha. It was used on tramways in Chicago, but the 1892–93 trials in Deptford and Croydon were not considered successful.

increased for grades, etc., or the car reversed. This mechanism is described and illustrated in the *Gas Light Journal* for Sept. 17th, 1888, and also in the Company's catalogue. The absolute superiority existing in a self-contained and cheaply operated motor for each car, rendering it independent of slots, grips, cables, conduits, trolley wires, batteries, or break downs of a central plant, presents such attractiveness, that, to the writer, it would seem that the gas motor is bound to grow into use.'

However, there is scant evidence that the vehicle described in either of these patents was ever built and by 1889 Connelly had started experimenting with a separate power unit to haul former horse cars, using an engine running on naphtha gas, contained in a large overhead tank above the vertical single-cylinder – and later two-cylinder – engine.

So it would seem that his belief that he could power a tramcar with a small lightweight engine and gearbox had not proved successful.

Connelly and his brother Thomas sought patents for numerous tramway vehicles, and both men also invented and patented governors especially designed to meet the requirements of gas engines.

On 3 July 1894, a detailed account of the first public run of the Lührig vehicle in Croydon was carried by several newspapers. The report in the *Cork Constitution* points toward the preparations for a fully functioning gas tramway being much further advanced than just a brief time-limited trial as has generally been assumed.

If the level of commitment to gas traction in the following extract is correct, then something must have gone seriously wrong for it to be so quickly abandoned:

'The charging-station is at the depot of the Croydon and Thornton-Heath Tramways Company at Thornton-Heath. where an 8-horse power Otto gas engine drives a compressor. By this latter the gas, which is taken from the gas company's mains, is pumped into a steel cylindrical receiver 25ft long and 4ft diameter, at a final pressure of ten atmospheres maximum, or 150lb per square inch. This plant is equal to the supply of five tramcars, the number which it is intended to place on the line at first. Five more are be added in due course, when another compressor and receiver will be put down and will be driven by the present engine, which is of sufficient power for the purpose. From the receiver a pipe is laid down to the tramcar charging-point, the cylinders on the car being charged in the same way of those of railway carriages are charged with gas for lighting purposes. The pressure in the tramcar cylinders is about eight atmospheres, or 120lb per square inch, at starting, and the cost of the gas used is stated to be 1d per mile with a fully loaded car. The trial, which took place yesterday at the insistence of the Traction Syndicate (Limited), was conducted without hitch or hindrance of any kind. The car is going into regular traffic at once, and the other four, completing the first set of five, are shortly to follow.'

These new cars were not going to be of the design referred to by Major-General Hutchinson. They were going to be small double-decked vehicles very similar in profile to the horse trams which were already familiar sights on British tramways.

The very positive picture painted by the newspaper coverage conflicted somewhat with a report of the proceedings of Croydon Council which appeared in the 7 July 1894 issue of the *Croydon Chronicle and East Surrey Advertiser*. Several objections to the introduction of the gas cars had been raised in council meetings, mainly on concerns about safety.

> 'In accordance with the recommendation of the last report of the Committee, the Town Clerk informed the Tramways Company that the Council objected to the tramlines being paved with granite setts.'

As in other places, replacing horse trams with gas or steam trams required well-ballasted and heavier-gauge rails, and granite setts were an obvious solution. The report continued:

> 'Read letter from the Board of Trade, enclosing regulations that they have made with regard to the use of the gas tram motor car. The Board have not adopted the suggestion of the Council that the licence should be terminable by the Council, but have reserved the power to revoke it at any time. The licence limits the use of the motor to the portion of tramways on which the use of the Connoly [sic!] motor was sanctioned, and the Board of Trade forward copies of regulations they have made concerning the same... As to the motor they thought the Council should have the power to revoke the licence at any time, but the Board of Trade had reserved that power. There was one thing in connection with this matter which the Committee had strenuously opposed, and that was that the gas motor should not be allowed to run to the "Crown" corner, no further than Poplar-walk. They were pressed to concede that point, but the Committee felt owing to the narrowness of North-end, it would be very dangerous, consequently they remained firm in their opposition.'

The next mention of the Croydon experiment in *The Engineer* was in its issue for 28 July 1899 where it was noted that 'It is now several years since gas trams were run for some time on the Croydon Tramways.' That article noted that the first was a relatively small car, fitted with a 7hp engine, very similar to those still being operated in Dessau, while a second, larger, double-decked British-built car had been introduced onto the line in 1894. That was incorrect – the Dessau car was only allowed one short run in 1893 before it was refused permission to operate on safety grounds.

The larger single-decked Dresden vehicle was the one licensed for the six-month trial, probably accompanied by the first of the double-deckers built by the Ashbury Railway Carriage and Iron Company.

However the statement by the Consulting Engineer Professor Alexander Kennedy that 'It is intended that the gas cars shall entirely supersede the horse cars on the Croydon and Thornton Heath Tramway Company's system' proved optimistic.

Had the British Gas Traction Company gone ahead and ordered its first five tramcars from the Ashbury Railway Carriage and Iron Company before it had finally been given approval by Croydon Council to operate the service? Researches so far have failed to answer that question.

By the time the 1899 article appeared, however, the tramway company was already erecting overhead wires for its new electric network which went live in 1901, so the six-month experiment had obviously not succeeded as anticipated.

Efforts to discover whether or not the order for five double-decked cars was ever completed by the Ashbury company have so far also drawn a blank – perhaps they were eventually assigned to the new Blackpool line.

Below: 'Sectional Elevation of a Lührig Gas-Motor Car' as published in *Cassier's Magazine* June 1895, showing the motor positioning and two gas tanks either side of the very short wheelbase on this Lührig prototype double-decked tram. This design had a longitudinal bench seat on the upper deck, with the cooling water tank beneath it.

Opposite top: The rebuilt Neath tram, standing outside its shed at Cefn Coed Colliery Museum with the recently restored and re-erected colliery headgear behind. In its original form, the end dashboards would have looked quite different. They would not have projected as far below the bottom of the chassis and there would have been metal 'cow-catchers' to protect the gas tanks and to stop anything getting trapped beneath the car. (*See* illustration of the same design of Ashbury-built car on pages 126–127.) Matt Price acquired the tram in the 1920s before selling it on for conversion into a garage.

Opposite bottom: The tramcar with its engine inspection doors open. While the engine was lost many decades ago, a replica of the heavy flywheel which helped keep the engine running smoothly is *in situ*. As the rebuilt tramcar sits on a repurposed two-axle former railway guard's van subframe, the wheels are about one fifth larger than would have been the case when the vehicle was in service. Originally, the body of the vehicle would have sat on much smaller and lighter coil springs than those seen here.

Between those two reports, however, much had been written about gas trams elsewhere, drawing attention to their economy and ease of operation.

Formed in 1893 as a partnership between The Traction Syndicate Ltd and the Lührig company, The Gas Traction Company Ltd acquired the exclusive rights for more than twenty countries, and set about redesigning the tramcar and improving its performance. The British patent rights would be transferred in 1896 to a new company – The British Gas Traction Company Limited – whose Chief Engineer was Lührig's co-inventor Lucien Alphonse Legros.

By 1894, three years after Lührig's death, The Gas Traction Company Ltd had developed the single-engined vehicle which would be tested at Croydon that year. In this design, the engine was sited beneath one bench seat, with a large gas bag encased in a steel cylinder beneath the other, replacing Lührig's earlier roof-mounted gas tanks.

Two other bags of gas were housed in rigid cylindrical casings beneath the floor. Lührig's first British patent (No.15,841 of 1892) had specified such a car.

The 'caoutchouc bags' he specified were made of thick and heavy latex rubber, designed to collapse under their own weight as the gas was used up, thus maintaining the required pressure.

They were also quick and simple to refill, using a rubber hose from a convenient gas tap at the depot. The limited elasticity of the bags, and the constraining effect of the metal containers within which they were housed, ensured they could not easily be over-filled.

The company also developed a double-decked tram – based on the style of the horse-drawn vehicles which were already familiar sights across Britain, and the first of these larger twenty-six-seater vehicles – fitted with a more powerful Otto-cycle engine – was also tried out, but Croydon eventually opted for electric trams.

This was a period when several towns were being courted by The Gas Traction Company, provoking some quite heated debates in council chambers.

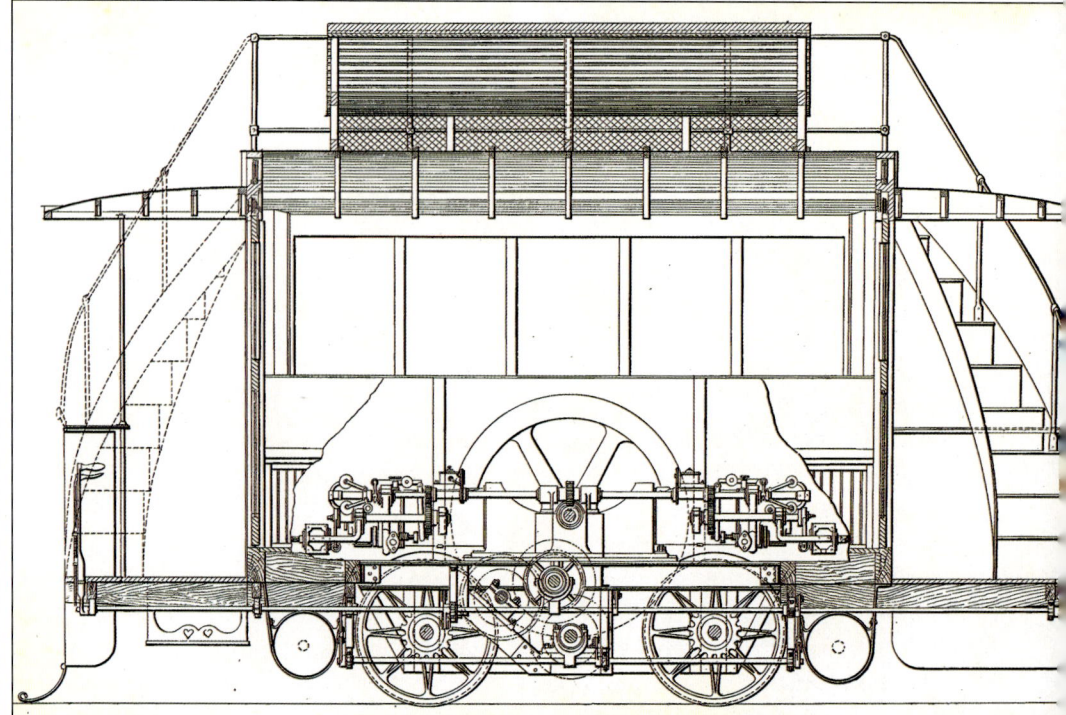

The 26 October 1895 edition of the *Rhyl Record and Advertiser* reported on the culmination of just such a heated debate in the North Wales resort where the die-hard 'trams will ruin our town' faction eventually came out victorious:

'The adjourned meeting of ratepayers to consider the advisability of sanctioning or otherwise a scheme of gas or electric tramways for Rhyl was held at the Board Room, Town Hall, on Tuesday evening. Captain Keatinge, J.P., presided over a large attendance.

The Town Clerk (Mr A. Rowlands) read a report of what took place at the last meeting when a proposal was submitted in favour of the gas traction scheme recommended by the Council, and an amendment that sanction be given to a scheme of electric tramways.

The Chairman having explained why the Road Committee of the Council had come to the conclusion that the gas traction system was preferable to the electric system said that personally he was very much in favour of the proposal.'

In the heated debate which followed – including some wild statements of what we would describe today as 'spin' – those against the scheme suggested that tourists would be killed by the trams and holidaymakers would stop coming to the town, boarding houses would go out of business, and that the £20,000 cost would be money wasted. It was never put to the test. Assurances that all construction costs would be met by the company fell on deaf ears, and in the end, not only was the original gas tram system rejected by the ratepayers, so was the electric alternative.

Progress was, however, being made elsewhere, and in 1896 the Ashbury Railway Carriage & Iron Company supplied four forty-seater cars fitted with 14hp German-built Deutz engines for use on the Blackpool, St. Anne's and Lytham Tramway operated by The British Gas Traction Company. As the line ran between two boroughs, Blackpool Corporation was contracted to supply the gas from their municipal gasworks at their end and, once the line was extended to Lytham, Lytham Corporation supplied it at the other.

By 1897 the company had purchased a fleet of larger fifty-two-seaters built by the Lancaster Railway Carriage and Wagon Company and fitted with larger Otto-cycle

Opposite: The interior of the restored tramcar. In this view, the engine would be beneath the seats on the left, a gas tank beneath those on the right. The overhead pipes would have fed gas to the interior lights.

Below left: On the engine side, the seat backs are set further in to the vehicle to make space for the flywheel which is located in the space between the seats and the outer body of the tramcar. If the engine was not set up correctly, passengers on this side of the vehicle complained of vibration.

Below right: On the gas tank side of the vehicle, the seat backs are set closer to the tramcar's body frame.

engines, built by Deutz according to Henry Holt's improved design. Several sources – none of them contemporary with the vehicles – suggest they were actually built by Crossleys, who had long held the UK rights to the Otto-cycle engine and had been co-patentees with Nicolaus Otto himself in several 'improvement' to the engine design, but the Crossley archives do not hold any records of the company ever building the horizontally opposed two-cylinder engine used in the tramcars.

It can be reasonably assumed, therefore, that throughout the period of gas tram operations in Britain, the engines continued to be imported from Deutz – and possibly serviced by their engineers.

Indeed, in his 1896 lecture to the Société des Ingenieurs Civil de France, published in the *Société's Mémoires et Compte Rendus des Travaux*, M. A. Lavezzari reported that:

> 'This line, the first regularly established in England, was only put into service last May, which means that its operation cannot yet give clear results. It goes from Blackpool-Saint-Anns to Lytham; its length is 13 km.
>
> The cars are of the larger model with an upper deck and can contain 40 travelers; their weight is 7 tons empty and 10 tons loaded. There will be sixteen cars in full operation.
>
> While in Manchester in February I saw the first three cars being built at the workshops of the Ashbury Railway Carriage and Iron Company, which is building the equipment for the Gas Traction Company, except the engines which come from Germany.
>
> I owe my thanks to Mr. Walter Gatwood, Engineer of Ashbury, to have seen in detail these cars as well as another which was returning from Paris and which was going to remain there.'

The line from Blackpool to St. Anne's on Sea had officially been opened on 11 July 1896 from Squires Gate in Blackpool to St. Anne's – not May as Lavezzari had said. According to a report in *The Belfast Newsletter* on 13 July 1896:

> 'A luncheon was held during the afternoon at the Clifton Arms Hotel, Lytham, over which Mr. Alderman Pilling. ex-Mayor of Southport, presided, and was supported by representatives of the surrounding Corporations. The chairman said the tramway was a great advance in civilisation. Blackpool, he said, would take St. Annes by the hand, and St. Annes would grasp Lytham, and there would be houses from one end to the other, and some day, he, said, the estuary night be spanned, and Lytham and Southport connected by a bridge. Colonel Ellis, who proposed the toast of "The Tramway Company," said the system was the best in the country. The other speakers included the Mayor of Blackpool, Mr. Fletcher Moulton Q.C., Mr. Roger Wallace, and Dr. Farrell.'

Alderman Pilling's prediction about the Ribble Estuary being bridged looked at one point as if it might be about to come true. Just two years after the Blackpool, St. Anne's and Lytham service was inaugurated, he had proposed building a bridge and tramway from Southport to Lytham, the estuary itself being crossed by a transporter bridge between North Meols and Lytham.

A Bill was laid before Parliament in 1898, modified in 1899 and a revised version resubmitted as *The Southport and Lytham Tramroad Act* and that version was passed in 1904.

It envisaged a tramway, powered by 'suitable machinery' – which, given Pilling's enthusiasm for the Blackpool system, might well have been gas engines. All that was specified in the Act was the use of 'suitable machinery'.

Although the bridge was never built, the Southport & Lytham Tramroad Company was not dissolved until 1936.

The section of track from Squires Gate to Station Road was owned by Blackpool Corporation, the remainder by the Gas Traction Company. In 1897, and shortly after the route was transferred to the British Gas Traction Company, the line was extended to Lytham, giving it a distance of around six miles, quite a bit shorter than the eight miles (thirteen km) Lavezzari had reported.

The directors of the new company were listed in the *St. James's Gazette* newspaper in its issue of 21 July 1896 together with some of the other company directorships which they held. They were listed as:

'Lt.-Colonel W. T. Ellis, Director, The Incandescent Gas Light Co., Limited. Thomas Skarratt Hall, Esq., 34, Berkeley-Square. W. Henry P. Holt, Esq., (Crossley Brothers Limited), Manchester and London. Thomas G. Gillespie, Esq., Director, The London Tramways Co., Limited. Alfred De Cros Esq., Dunlop Pneumatic Tyre Co., Limited.'

In that same issue, the prospectus – and letters of support and commendation – for the new company declared that:

'The Company has been formed for the purposes of acquiring, working, developing, or otherwise dealing with the British patent rights, including all improvements thereon, for Great Britain and Ireland for the system of working tramways, light railways, etc., by means of motor cars worked by gas engines, invented by the late Carl Luhrig, of Dresden, and Mr. Henry P. Holt.'

It was first referred to as the 'Lührig-Holt system', in the prospectus published at the time the Gas Traction Company was registered in December 1893, recognising the important improvements to both the engine and Lührig's tramcar itself, which had been introduced by Henry Holt. The joint name was repeated in the 1896 prospectus for The British Gas Traction Company by which time Henry Holt was also a director of both Crossley Brothers and the new company.

A major limitation with all these vehicles was the amount of gas which they could carry in their three gas bags – enough to travel between twelve and sixteen miles which meant that they had to return to the depot after every second trip to refuel with pressurised gas at around 8 bars – supplied via a 4hp Crossley gas engine, of course.

Initially there was only one 'Charging Station' – at the company's Squires Gate depot in Blackpool – but in 1897 with more trams on the extended line, another was opened at Lytham's Market Square depot to speed up turnaround times at either terminus.

Despite the fact that most people used gas for cooking and gas mantles for lighting, there were reports of public concern over the safety of the gas cars, so the prospectus included some statistics to allay whatever fears investors might harbour:

'As evidence of the safety with which gas traction can be worked, it may be stated that during the years 1894–1895, there were at Dresden 212 accidents with horse cars, 88 accidents with electric cars, but not a single accident with Gas Motor Cars.'

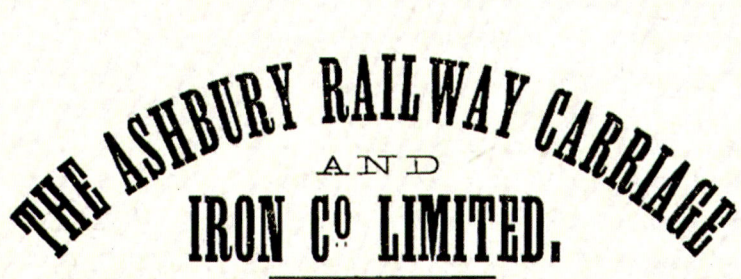

Opposite: Published in *The Engineer*, 28 July 1899, a detailed engraving of the rear view of the Blackpool, St. Anne's and Lytham Tramway Company's gas tram No.1 – looking similar to the preserved Neath Tramways Car No.1 – sadly now missing its 'cow-catchers'. Note the 'garden bench' seating on the top deck. While the lower saloon of the tram was lit by gaslight – of course – the only illumination the driver had at night was a small oil lamp. Three years before it was published in *The Engineer*, the same illustration had already been published in *Supplement No.1077* published with the *Scientific American* on 22 August 1896

Above: An advertisement for the Ashbury company which built the bodywork for Blackpool's first gas trams.

That was a typical marketeer's distortion of statistics in order to prove a point – Dresden had over 120 trams, the majority of which were horse-drawn, many were electric, and only five were fuelled by gas so the absence of any serious accidents was not representative of their safety when compared to other vehicles.

Percy Holyoake, Secretary to the British Gas Traction Company, very quickly became aware that unless the system was adopted widely, the company was likely to

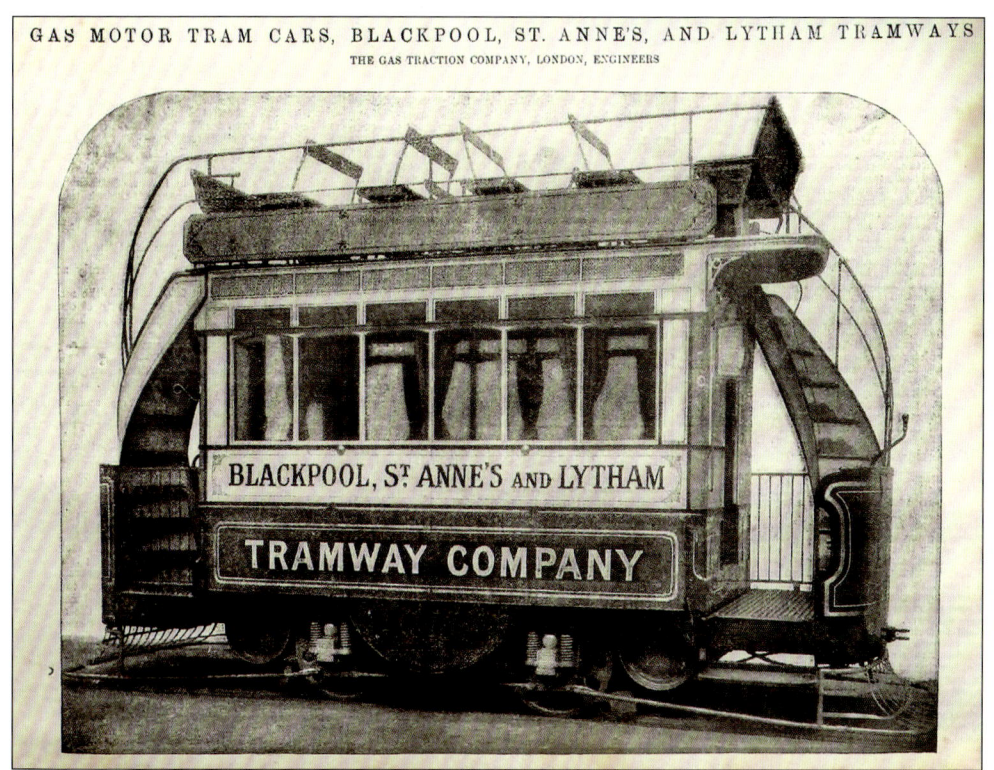

From the journal *The Engineer*, 28 July 1899, a photograph of one of the original small cars from the Blackpool, St. Anne's and Lytham Tramway Company's fleet of gas trams.

The company's name seen here on Car No.1 – Blackpool, St. Anne's & Lytham Tramways Co. Ltd. – differs from the *The Engineer* illustration on the previous page. The use of 'Tramways' rather than 'Tramway' with an ampersand in place of 'and' dates from after the British Gas Traction Company had relinquished the route to the new company in 1898. Only two photographs of this particular short-lived livery have been seen during the research for this book, both showing Car No.1.

From the same period, another of the small tramcars at the height of the summer season – as one group of passengers disembarks, another group waits to board.

quickly run out of money, so he set about trying to persuade a number of towns and cities to adopt gas propulsion and to encourage newspapers to report on the success of the system. Like any good marketeer, he did not hold back in his eulogising about its success, and in seeking to persuade other towns and cities to sign up for what he genuinely believed was the traction system of the future. However, few agreed with him and, according to a report in the *Westminster Gazette* on Friday 28 October 1898, 'such corporations as Leeds, Belfast, Glasgow, Sheffield, Birkenhead, and Liverpool,' had 'after thorough investigation. all negatived this very system.'

A few months earlier, the *Northern Whig* newspaper had reported on 25 March 1896 that a deputation of council officials and experts from Belfast had visited several towns and cities in Britain and mainland Europe to get a better idea of the

THE GAS TRAMCAR IN BRITAIN • 129

Blackpool, St. Anne's & Lytham Tramways Car No.18 – one of the fifty-two-seater trams built by the Lancaster Railway Carriage and Wagon Company – reproduced from *A Guide to Blackpool* c.1897 when the terminus was still at St Anne's.

From *The Engineer*, 28 July 1899, a detailed drawing of the mechanics of the Blackpool, St. Anne's & Lytham Tramway Company's Car No.1.

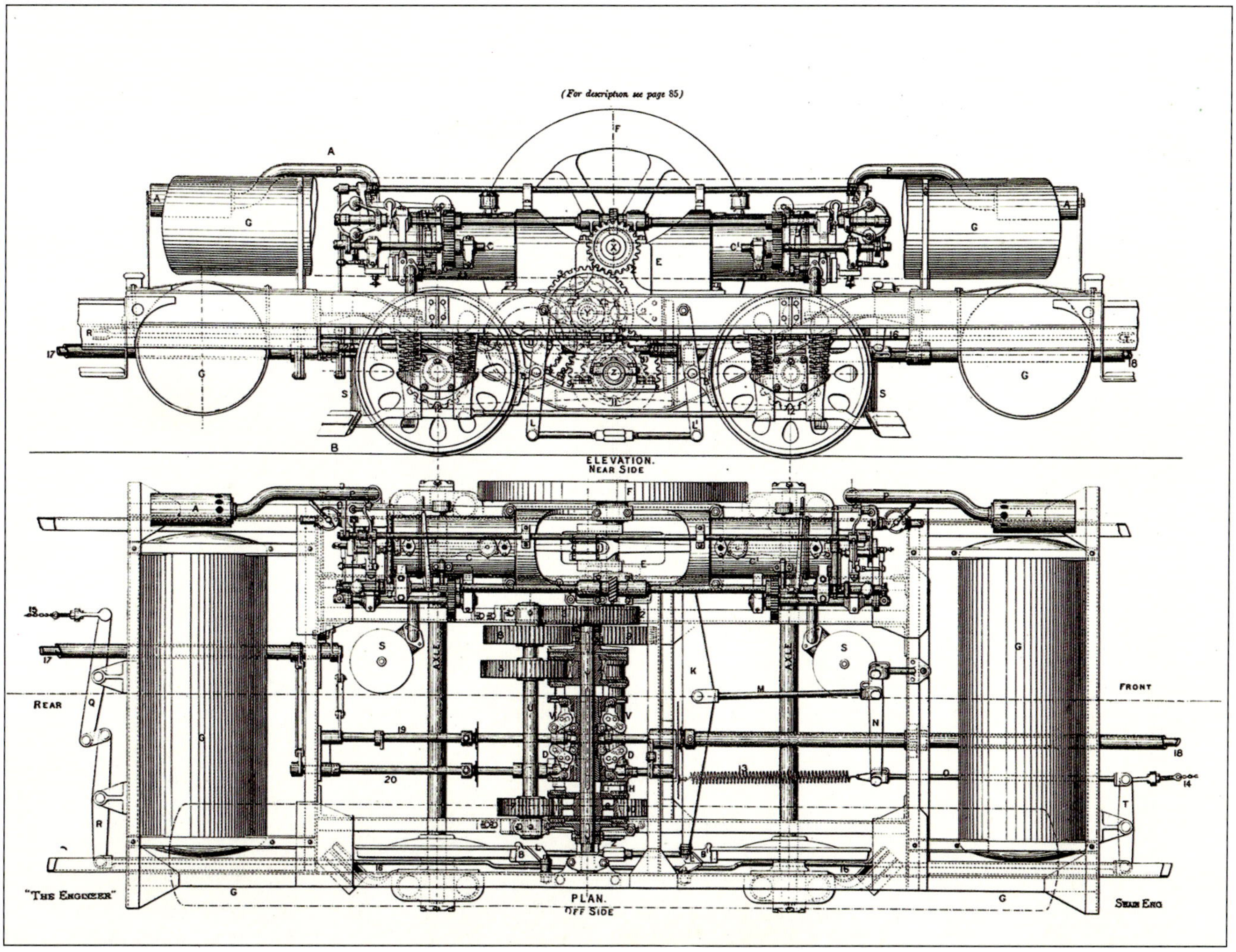

132 • THE GAS TRAMCAR

Previous pages: Another of the large Lancaster-built trams, supplied to British Gas Traction for the Blackpool, St. Anne's and Lytham Tramways Company. This vehicle does not carry the company name on the strip beneath the windows, but the crest on the flywheel doors dates the photograph to 1896–97 and the period of ownership by the British Gas Traction Company. After the sale of the line and tramcars to the newly established Blackpool, St. Anne's and Lytham Tramways Limited in 1898, the previous operator's crest was removed from the trams. The size of the underfloor gas tanks on these larger trams can be appreciated in this photograph.

Right: Car No.16 was one of the fleet of larger and more powerful tramcars built for Blackpool, St. Anne's and Lytham Tramways Limited by the Lancaster Railway Carriage and Wagon Company in 1897 and fitted with the more powerful 15hp Deutz engines.

traction systems then available. Their report, however, did not endorse Holyoake's enthusiasm, claiming that:

'The objectionable features in connection with steam or gas as a motive power are too great to allow of their being recommended in any known form.'

The Engineer reported on 14 May 1897 that a deputation of council and tramway officials from Sheffield had undertaken a similar fact-finding tour of every sort of tramway – including the Lytham gas-engined line – to assess which would be best suited to their somewhat hilly terrain. The report they produced listed their several concerns, noting that:

'Although a long way behind its competitors in every respect, gas as a motor in tramway traction has several advantages which ought not to be either ignored nor undervalued. Except at Blackpool, the deputation saw no instance of tramways actually working by gas power. It is true, it visited Dessau, where there is a system of that description, but the members were not gratified with witnessing it in actual operation. As this line was opened for traffic more than two years ago, some

The Gas Traction Company Limited issued £100,000 of share capital, available as £1 shares in any number from one upwards and was incorporated on 24 November 1893. The certificates were printed by Waterlow & Sons, the well-known London-based engraver of banknotes, postage stamps, stocks and share certificates. The tramcar illustrated on the share certificate is of a design which does not seem to have ever been put into production. It is of a design illustrated in Lührig's 1892 British Patent No.15,841. The Company Secretary was Percy Holyoake, and Henry Holt was one of the shareholders. Both would later become shareholders in the British Gas Traction Company Limited as well – as did Thomas Gillespie of the London Tramways Company, although London never adopted gas trams. Its successor, the British Gas Traction Company, raised £250,000 in share capital.

explanation why it was not working at the time the deputation visited the town might have been given in the report before us. We shall return to Dessau shortly, as it merits a brief description to itself. It is admitted that at Blackpool the gas tramway is on its experimental probation, and a length of seven miles is a very fair stretch for experimental trips. The chief objections mentioned in the report, to the gas tramcars are the vibration, the unpleasant smell, and the great difficulty in ascending gradients; but, per contra, they require no expensive installations, neither subterranean conduit nor overhead wires, the cost of traction is comparatively low, and they are easily charged at the depot.'

That provoked a strongly-worded reaction in the following issue from those involved in operating the Blackpool, St. Anne's & Lytham line. They particularly rejected the idea that theirs might still be considered to be an experimental tramway and that the use of gas traction was still operating on probation.

'Sir. In your issue of the 14th inst., page 480, reference is made to some observations of the Sheffield deputation which might mislead the public as regards the present and future of gas traction in this country, The deputation appear to have made their observations on the earliest experimental cars, as our tramcars at Blackpool are practically free from vibration when standing and go up the gradients of about 1 in 26 with ease. The change of speed referred to when descending this gradient is due merely to the change of gear provided for the purpose, the actual power of hill-climbing being increased with the observed decrease of speed. The cars on the Blackpool line have frequently been loaded to about double their capacity, and have climbed the gradients in question with this load. From 31st July last they have run 31,500 miles. The line and its stock are in no way an experiment, they have accomplished from the first all that was expected of them, and have, in fact exceeded expectations. The Blackpool, St. Anne's, and Lytham Tramways Company, are moreover so far satisfied with the system that they have, now that the whole length of the line is completed, ordered twelve more cars of similar construction to complete the equipment of their system.'

Referring to the Sheffield delegation's claim that none of the Dessau cars had been running when they visited, the writer suggested that perhaps they had gone to Dresden instead of Dessau, pointing out that:

'…at Dessau the system has been working continuously since November, 1894, with a large saving of cost as compared with that of haulage by the other systems. At Dresden a line was worked for about a year, 1893–4, and will be worked again when the new line is completed.'

Early that month Holyoake was in Ipswich, promoting the system, taking with him a model of a tramcar and a mock-up illustration, based on one of the small Blackpool, St. Anne's and Lytham cars. That was published alongside an interview with him in *The Evening Star* on 3 May. The interviewer – who had probably never heard of a gas-engined tramcar before, and had only just been shown a model of it when embarking

Car No.19, fully laden on Church Road, Lytham. Despite their size, these large trams still had a very short wheelbase, causing one disgruntled passenger to describe the experience of travelling on the top deck as 'like a choppy sea voyage'. No.19 was one of those which would be sold to Neath.

The Drive, St Annes on the Sea. 22-7-8bm.
En route pour Buxton!

A 'Reliable Series' postcard of The Drive at St. Anne's on the Sea, photographed c.1902, with a large number of elegantly clad Edwardian holidaymakers waiting to board gas tramcar No.12.

on his assignment – was sceptical about the system, suggesting the public would not want to travel sitting on a cylinder of compressed gas, and would be terrified of a possible explosion. His questioning barely concealed his conviction that such vehicles were inherently unsafe.

As well as pointing out that railway carriages had carried compressed gas cylinders for lighting for more than twenty years without incident, Holyoake mounted a vigorous defence of the system, adding some interesting information about its development:

> 'Our first experiments in this country were made at Croydon, during the night and early morning, and none of the horses drawing loaded waggons to Covent Garden, with the drivers sometimes fast asleep, ever took fright at the gas-traction cars.'

In the *Evening Star* article, Holyoake paid tribute to 'the inventive genius of Mr. H. P. Holt of Messrs. Crossley Brothers' for the fact that noise and smell 'have been practically extinguished by the use of special air and exhaust silencers' while special lubricants had considerably reduced the vibration which had beset early gas cars.

This repeated reference to Holt being 'of Messrs. Crossley Brothers' is a curious one. Holt, as has been explained, was also a director of the British Gas Traction Company, and associating his name with that company might have seemed more logical if any records could be located to suggest that Crossleys had been in any way involved in the construction or maintenance of the engines, but they don't. So we must assume that the company was simply cashing in on Crossleys' reputation.

When asked by the journalist 'What is the practical outcome of your great experiment at Blackpool?', Holyoake replied:

'Our success has equalled the highest expectations. You may make a note of the facts that since last July we have run 40,000 miles there, and never once had a breakdown, that twelve more cars have been ordered for the coming season and will shortly be put in traffic, so well do the public like the service. The peculiar circumstances under which we started at Blackpool and are now working have very happily served to bring out the special advantages of the gas-traction system. Before our advent an electrical tramway was already in existence; after an infinite amount of trouble we got powers to carry on an extension line to Lytham and St. Ann's (sic), and so the two kinds of motor power were brought into direct comparison. I don't hesitate to say that the gas-motor has conspicuously triumphed. Each one of our cars, you see, is self-contained and independent. If one of them should break down from any cause, the rest go on running just the same. With electrical tramways however – and cable tramways too – the case is different. A breakdown with them means the break-down of the whole system, and so it happened that some time ago, when the Blackpool electric tramway was stopped owing to storm and floods, our cars were running as merrily as ever… On all accounts, gas-traction is going to hold the field.'

Had the system been developed a few years later when more powerful engines were available, Holyoake might have been proved correct, but despite all the setbacks, his advocacy of gas traction had been met with some enthusiastic responses, even if they were never ultimately seen through to completion. Writing in the *Westminster Gazette* on 22 June 1899, Thomas W. Hersey,

Above: Car No.16 in Clifton Square, Lytham, the terminus of the route.

Opposite: Public fascination with the idea of gas tramcars was widespread, albeit tempered with a little uncertainty over their safety. This illustration of what an Ipswich gas tram might look like accompanied the interview with Percy Holyoake in the *Ipswich Evening Star*. Holyoake promised the newspaper 'The public generally will be little concerned with mechanical details, the less so because nothing will be seen of the internal arrangement. Nothing will be seen or smelt, and next to nothing heard.' Ipswich did not adopt the system, so his claims were never tested.

a former employee of the British Gas Traction Company, reminded readers that the previous year:

'London County Council's Engineer reported on the workings of the system at Blackpool, with the result that the Highways Committee of the Council intimated their willingness to consider a proposal for a trial of the gas motor cars of the British Gas Traction Company.'

He then launched into an attack on what he saw as the inactivity of the Board of the company, berating them for missing an opportunity to put gas traction formally on the map.

'Sir, considering the large amount of money that has been expended to demonstrate the efficiency of gas traction in this country and in Germany, it is nothing less than astounding that the managing committee responsible for the success of gas traction (which consists of Mr. J. Fletcher Moulton, Q.C., M.P., Mr. Roger W. Wallace, Q.C., and Mr. H. Holt, C. E., of Crossley Brothers and Co., Limited) apparently do nothing to convince the public authorities all over the country, and especially in London (seeing that the County Council are on the eve of arriving at a momentous decision as to the best mode of traction for the tram lines they have taken over) of the practicability of using gas as a motive power for tramways. I can assure you that gas and other engineers consider that gas traction is dead. During the last session of the Institution of Civil Engineers it was openly stated that nothing was being done, so far as the public knew, to show that the system was being put to the front at this crisis in London and elsewhere; and only the other day one of the most eminent engineers in

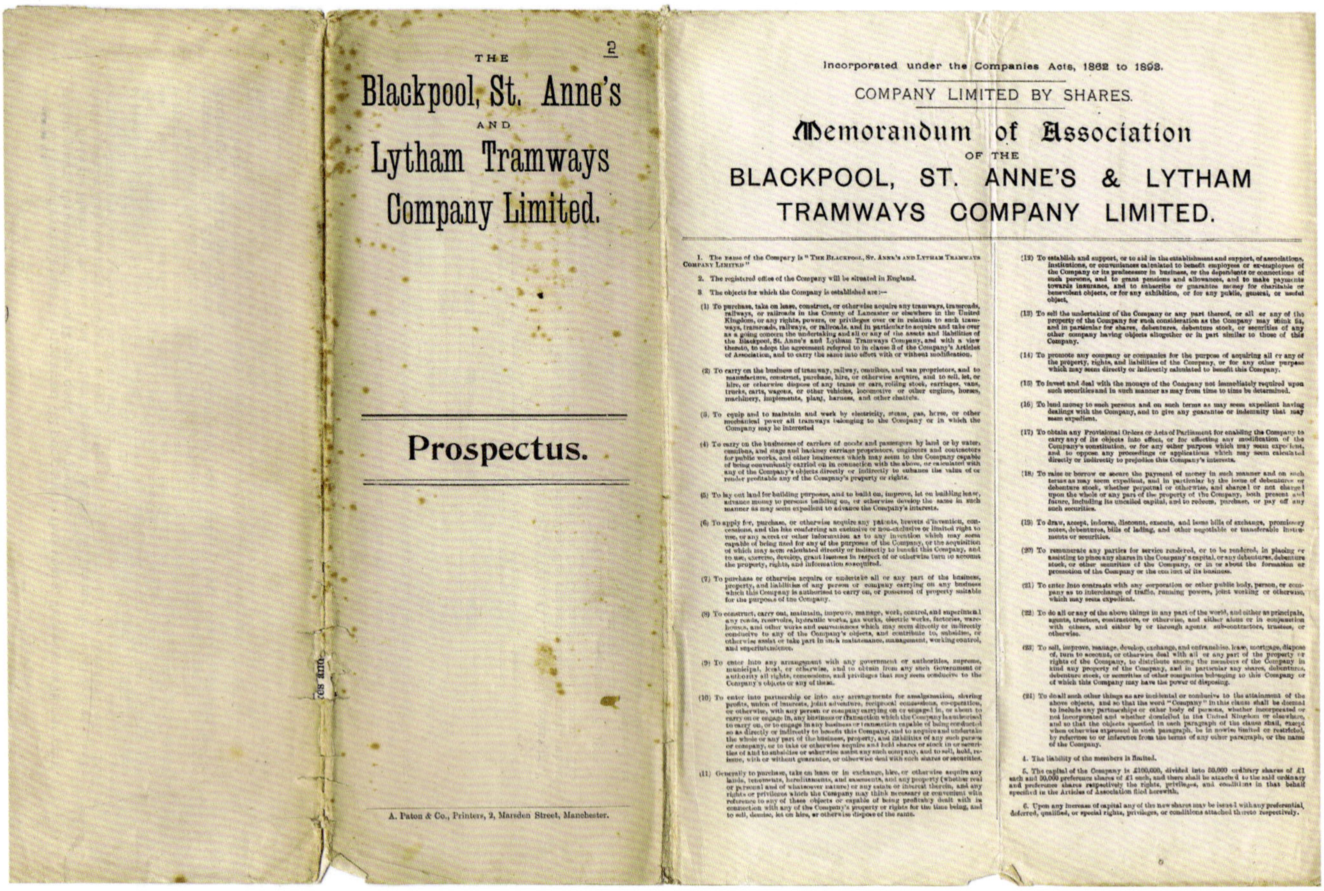

London remarked to me that when once the London County Council decided upon adopting electricity the chance of introducing gas traction would certainly be lost...'

He was correct – London was the key. With London lost, gas cars would never be developed beyond their current under-powered state.

There were other proposals – on 28 October 1898 the *Westminster Gazette* had reported the British Gas Traction Company's Chief Engineer Lucien Legros saying that the company 'is now constructing a line from Bideford to Westward Ho – to be worked as a light railway', but in December the following year, the *Harland and West Country Chronicle* reported that the directors of the Bideford and Westward Ho! Railway distanced themselves from the project, despite the fact that they had already invested in the scheme. The Light Railway opened, steam-hauled, in 1901.

Less than a month after Legros's statement, a report by London County Council's 'expert' J. Allen Baker was published on 20 November – again in the *Westminster Gazette* – comparing gas traction unfavourably with either an American compressed air system, or with electrical conduit.

While the argument made some appropriate points, it limited itself to operating costs, ignoring the huge capital investments needed by either of those alternative systems:

'Were I convinced that it is advisable to equip our London tramway system with separate motor cars, my preference would be for the improved compressed air system, of the American Air Power Company of New York, as opposed to the gas system. However the smell of the gas trams be disguised, the result is always a befouling of the atmosphere, and I ask you to consider, Mr. Editor, the effect of a thousand or more trams on this system constantly emitting the fumes of burnt gas into the already too impure air we Londoners breathe.'

How he would have responded to London traffic today cannot be imagined, and he concluded:

'I might further add for their information that the Blackpool gas trams run from an outskirt of that town through a country district (and for such they may be suitable) to St. Anne's and Lytham. In my report I was not dealing with country districts, but with London.'

As it was, the British Gas Traction Company, having failed both to get the London trial underway, or to get Blackpool Corporation to allow gas cars to run over its lines, decided that it would be impossible for them to run the line economically and, in October 1898, they had transferred their rights to the Blackpool, St. Anne's & Lytham Tramways Company Ltd, selling all their assets to them for a reported £115,000.

That may have been a somewhat premature decision, for in October 1900, the General Manager of the Blackpool company approached his counterpart in Trafford Park in an attempt to purchase their trams. Just a few months earlier, ownership of the Trafford Park line had passed from British Gas Traction to the West Manchester Light Railway Company, bringing railway and tram movements on the network under single ownership.

He was turned down but, undaunted, tried again six months later to be met with the same outcome. So, despite some concerns being voiced about the power of the trams and the challenges of operating them, there was clearly no suggestion at that point that

Opposite above: The 1898 Prospectus for the sale of the Blackpool, St. Anne's & Lytham Tramways Company Limited, when the British Gas Traction Company decided to cease operations on the line. The assets offered for sale were valued at £125,000 – half the value of the company's original share capital – and included several large plots of land, the compressing and fuelling stations at Blackpool and Lytham, the tram sheds at Squires Gate, four small and sixteen large tramcars – with a combined value of £10,000 – and leasehold access to the route along which the tramway ran. The Prospectus also underlined the quality of the trackwork – 'Barrow steel rails of 92lb to the yard' laid on concrete foundations and paved with granite setts. Operational figures for the twenty weeks up to 15 September 1898 showed that a total of 512,954 passengers had been carried on the tramcars, with a projection that a further 130,000 would be carried before the summer season's end. On average throughout that period, ten cars had been operating daily, but that would reduce to four in the winter season.

Opposite below: Blackpool, St. Anne's & Lytham Car No.12 moving at speed, possibly along either Church Road near Seafield Road on the outskirts of Lytham, or along Ansdell Road – local opinions differ – and was probably taken c.1902.

Right: Cars Nos. 14 and 17 photographed at St. Anne's probably in 1902. Both these cars are believed to have survived the 1903 storm and been part of the fleet sold to Neath, remaining in service there until 1920.

Below: The opening paragraphs of the *Lytham St. Annes Express* story published in December 1935. The photograph used dated back to 1898, with what are thought to have been the company directors painted out (see page 116).

gas traction was unsatisfactory or soon to be abandoned in Blackpool – quite the reverse, in fact, with the company still seeking to increase its fleet.

Had Trafford been willing to sell, the trams would have had to be re-wheeled before they could be used on a street tramway – their wheels having been specially adapted to run on the conventional railway tracks at Trafford Park rather than Loubat-profiled tramway rails.

However, it was all short-lived – although it seems clear that the real reasons for the decision were financial rather than mechanical as have often been reported. Some sources suggest that the gas trams were already experiencing mechanical issues before the service was withdrawn, but the fact that some of those cars went on to operate in Neath in south Wales for almost twenty years would tend to undermine that suggestion.

As it was the only gas-powered route on its entire network, perhaps Blackpool simply sought uniformity of traction.

Only two other tramways had been persuaded to adopt gas cars, and a decision was made in late 1902 to start withdrawing the Blackpool fleet, the exemplar for gas-traction, after less than seven years' service – a decision presumably brought to the forefront of their thinking a few weeks later after at least a dozen tramcars and the Squire's Gate depot were destroyed in a storm on 27 February 1903.

That event, while passing barely reported at the time – although there was wide coverage of the gales which caused the damage – was brought to prominence thirty-two years later, when one of the drivers, a Mr. W. H. Cartmell, retiring after more than thirty-five years of service with the British Gas Traction Company and its successors, was the subject of an extensive report in the *Lytham St. Annes Express* on 13 December 1935.

Although contradictory in places, the article contains a valuable first-hand account of the challenges crews faced in maintaining gas tramcars, and operating them from open driving cabs in all weathers. For example, Cartmell recalled that despite the interior of the cars being lit by gas lamps – lower deck only, of course – all he as the driver had with which to reassure himself that the way ahead was clear and pedestrian-free was 'the feeble light of an oil lamp', thus making the journey between Clifton Drive North and Squire's Gate a potentially hazardous affair for both trams and pedestrians during the hours of darkness. The newspaper continued:

'For some years after entering the service of the Tramways Company he was employed in the sheds, which at that time were on the opposite side of the road to the present sheds. It was whilst he was engaged in this work that Mr. Cartmell and other workmen had an alarming and, as he describes it, amusing experience.

It was in the early morning of a winter's day in 1903. He and several other men went down to the sheds to get out the first car – a workmen's special. It was terribly windy, and walking in the teeth of the gale was a difficult matter. However, they arrived at the sheds and proceeded to open the great doors. Immediately they did so the wind howled into the building, lifted the roof off, and caused the whole structure to collapse, in spite of the fact that, like the present sheds, the outer walls were constructed of brick.

Mr. Cartmell smiled as he recalled the incident.

"No, I don't think anybody was hurt," he said in reply to my question, "but we were scared. Anyway, we got the workmen's tram away."'

The giant flywheel of a gas engine dominates this view of the wreckage after storms and gales destroyed the Blackpool, St. Anne's & Lytham Tramways' depot near Squire's Gate on 27 February 1903. The storm lifted the roof off the tram shed, which crashed back into the building and damaged or destroyed most of the fleet stored inside at the time. Two of the wrecked tramcars still carry advertisements for Blackpool's Winter Gardens on the stairways.

Two of the four 52-seater cars operated by the British Gas Traction Company on behalf of the Trafford Park Estates, photographed at their terminus. The cars are in pristine condition, but still painted in 'lead grey' primer instead of the tramway company's livery, suggesting they were new when the photograph was taken. When all four cars were eventually delivered, the fleet was painted in a green and cream livery with gold lining – see overleaf.

From newspaper accounts of the storm a dozen of the company's eighteen tramcars in the shed at the time were either totally or partially destroyed.

In an interesting anecdote, Cartmell recalled that while he worked for the Blackpool, St. Anne's & Lytham Tramways Company, he was not employed by them – only the conductors were paid by the company while the drivers were sub-contracted to drive the vehicles while employed by British Gas Traction. Why that administrative distinction between drivers and conductors was made, and what the accounting benefits to the two companies might have been, is difficult to understand.

Unlike other critics of the trams – and perhaps surprisingly given that he had driven the vehicles for several years – he made no criticism of their reported lack of power, but related some of the many other challenges of operating them.

'"There used to be some fun with the things" Mr. Cartmell said. "When we wanted to start up we turned a large flywheel, which was encased in the side of the tram about midway along it, and once we got started we chugged along very nicely. But there were lots of difficulties. The cars had a habit of leaving the track at curves. This was due to the fact that the wheel flanges were weak and used to break off with the result that when we were taking a bend the car, instead of going around it, would carry straight on and end up near the footpath.

Then the fun began. Gangs of men put fishplates down and tried to get the car back to the track. Sometimes it was done very quickly, but as often as not, it wasn't.

If the accident occurred at the bottom of a hill, such as at Squire's Gate, it used to take two and sometimes three other cars to drag the breakdown to the top. Oh, yes, we had plenty of fun – and plenty of real hard work."'

While Cartmell's account of his experiences as a driver had a light-hearted 'jaunty' feel to it, three years earlier on 9 September 1932, the same newspaper had published one disgruntled passenger's memories of what travelling on the trams really had been like.

Under the heading 'Riding Like a Choppy Sea Voyage', Mr. C. D. G. Hoare M.A., who described himself as a long-term resident in the town and local schoolmaster, was anything but charitable in his indictment of the trams – citing his local pedigree in support of his views:

'It must have been about 1896 that a momentous event occurred in St. Annes in the shape of the new trams starting their chequered career from Squire's Gate to Lytham. They were said to be the first Gas Trams (we had no electric light then) to be run anywhere, and I should hope they were the last. Running, as they did, on very short rails, and being considerably overhung, a journey on the top was rather like a choppy sea voyage minus the sea, while only those who were passionately fond of gas, ventured inside! St. Annes turned out to see the first ones – in which you could get a free trip – come hurtling down The Drive. In the light of after-events one can only regret that the idea of 'trams' ever materialised, both for the sake of share holders and ratepayers.'

Mr. Hoare's memory was a little distorted by time – and by what he deemed the unpleasant experience of travelling on the gas trams – but at a maximum speed of somewhere between 8 and 11mph, they can hardly have been 'hurtling down the Drive'.

Of course, the years of gas operation were underwritten by two private companies and not by the Lytham ratepayers. What he meant by 'short rails' is unclear – perhaps he intended to describe their 'short wheelbase', and the trams certainly needed that to help them cope with tight radius curves.

Interestingly, an article reporting the inauguration of the Blackpool and Lytham project, in *Supplement 1077* to the journal *Scientific American* published on 22 August 1896 had painted a much more positive picture of the system – one of which Holyoake would definitely have approved. That report also gave some interesting – if at times confusing – detail on the engine, sometimes describing it as a two-cylinder engine, and at other times as two, inter-connected, single-cylinder engines:

'The tramway is not yet open throughout its whole length, in consequence of sewer work obstructions at St. Anne's, but of the total of about seven miles of line owned by the company and line run over at Blackpool, the inaugural runs were made over nearly four miles. Three cars were running, and met the party of visitors near South Shore Station, Blackpool. All are fitted with precisely the same form of duplex Otto engine and the same form of gearing… The cars carry sixteen passengers inside and twenty-four outside, are very roomy and well finished, and weigh, with everything ready for the journey, 7½ tons. The engine is about 14 horse-power, and

A photograph taken shortly before the Trafford gas trams were withdrawn in 1908. This is believed to be Car No.3. Only four of the planned six cars were ever purchased and operated, and when the service was withdrawn, all four were sold at the June 1908 auction to a scrap merchant in nearby Barton, for just £80.

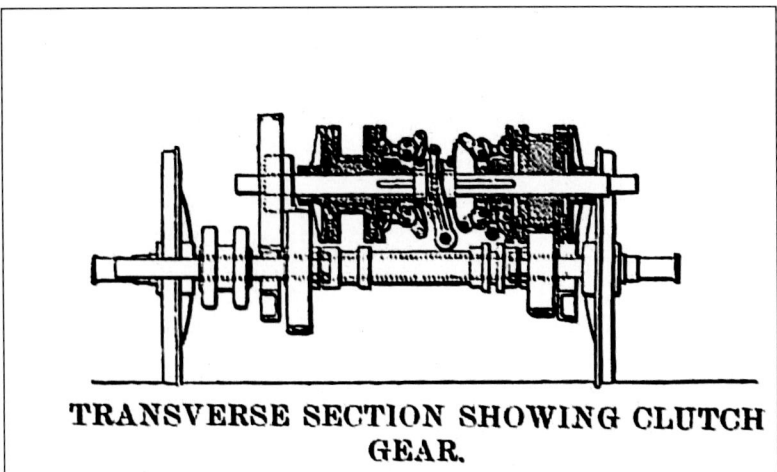

An illustration of the clutch gear on the Otto engine, as illustrated in Scientific American.

is placed on one side under the seats, and the gear is partly there and partly on the floor… The engine has two cylinders working on the one crank, the pistons being 7¼in. diameter and 9¾in. stroke. The space on the side opposite the engine is fitted with gas receivers, and there are two other receivers transverse to the cars… The water required for the cylinder jackets is carried in tubes on the roof, and circulation is maintained by by a smll pump, the water passing from one engine to the other in its circuit. The engines are balanced, one cylinder being opposite, or 180 degrees from the other… They are so arranged with regard to the gearing and governor that when on very easy roads only one of the two cylinders receives a charge, and thus less work is done and less gas used.

When stopping, not only is one cylinder cut out, but the speed of the engine is brought down from the ordinary 260 revolutions, subject to the governor, to about seventy-five revolutions per minute, the gas supply and the speed of the governor being altered by the movement of the hand lever, which throws the friction clutch of the driving gear out of action. At the same time the lubricating oil supply is reduced, so that no smell arises from the excess oil supply, which, under ordinary circumstances, collects when the engine runs light. This is effected by a very ingenious arrangement of the lubricator, by which its action is controlled by the taking of a charge of gas from the cylinder. No gas used, no oil used.

The engines and machinery are remarkably well balanced; so well that when the visitors approached and entered the cars, no one thought that the engines were at work. When the full gas supply is given to one or both of the cylinders, and the gearing put into action by means of the wood-faced clutches, some movement is appreciable, but not more than ordinarily noticeable with any car when running. There is a little "dither" caused by the new and unworked cog wheels, but this will soon wear off, and there will be nothing by which travelers can distinguish between these and ordinary cars, except speed and horselessness.'

The gas fleet was eventually replaced by electric cars in early 1903, the overhead wires which Percy Holyoake disliked so much having already been erected before the last gas car made its final run.

The British Gas Traction Company had also seen the potential of the development of the new Trafford Park Industrial Estate near Manchester, the first in the world, being built on land which the London investor Ernest Terah Hooley had bought from the De Trafford family for what was then the huge sum of £360,000.

That project had been stimulated by the opening of the Manchester Ship Canal in 1893, and within three years of the canal's inauguration, businesses were already planning to move their premises close to the canal and with space to expand their operations. The industrial estate was ideal.

However, it needed a transport system, and its owners determined that, for reasons both of safety and efficiency, they needed to be able to control traffic movements on that system themselves. Thus they decided against granting access to the site for Manchester's and Salford's trams and buses. Instead they would build their own

One of the Trafford gas tramcars alongside a goods train on the Industrial Estate's private railway, probably about 1900. Tramcars and railway locomotives and rolling stock using the same trackwork was the cause of several accidents during the gas tram era.

railway and tramway and sub-let the tramway operation to the British Gas Traction Company.

Tramcars would share the same rails as industrial locomotives from the Ship Canal Company – a decision which would turn out to have repercussions. They introduced four gas trams on the private railway on 23 July 1897, but the service was suspended within a few days when two female passengers – Elizabeth Kendrick and Sarah Timms – were hurt when their tram was derailed while negotiating a curve.

Clearly the wheel flange problems which driver Cartmell had recalled from his days with the Blackpool, St. Anne's & Lytham Tramways' cars were a problem on the Trafford Park line as well. But it was a surprising incident, for while the wheels might well have snagged while turning a corner on the conventional tramlines used at Blackpool, running on the tracks at Trafford Park should not have caused such a problem for, as the *Berkshire Chronicle* reported on 31 December 1898:

> 'This line differs from the one at Blackpool (which has ordinary grooved tramway lines) in the respect that it is a light railway laid with ordinary railway metals on sleepers, in order that goods engines and trucks may run over the line from the Ship Canal Docks.'

Although neither woman was seriously hurt, they threatened to sue the tramway operators – an inauspicious start for the new service – requiring Percy Holyoake to personally make the trip to Manchester to inspect the site of the accident for himself.

What he uncovered was a catalogue of failures in the way in which the trackwork had been designed and laid, creating a tramway which he considered to be potentially dangerous – and he was soon proved to be correct.

In 1896 the decision was made to order six tramcars from the Lancaster Carriage Works, although in the end only four were acquired. These were the same size as the larger cars being supplied to the Blackpool and Lytham line, and later to Neath Corporation Tramways, powered by engines from Gasmotoren-Fabrik Deutz.

NEXT WEDNESDAY, JUNE 3rd, 1908.

CAR SHED, TRAFFORD PARK, BARTON.

To Contractors, Machinery Brokers, Engineers and others

EDWIN BRADSHAW & SON

Have been instructed by The Trafford Park Estates Co., Ltd.,

TO SELL BY AUCTION
THE
GAS CARS,
ENGINES AND PLANT,

COMPRISING

Four Gas Cars, 12 B.H.P. Engines, by "Gas Motorenfabrik Deutz," to carry 52 passengers; Horizontal Steam Gas Compressor, 10-in. Cylinder, 12-in. stroke, with Waterjacketted Compressor Cylinder and 8-ft. x 4-ft. Vertical Boiler, by "Tangye, Ltd."; 2 Weldless Steel Gas Containers 20-ft. 6-in. x 4-ft. with Valves, etc.; 1 Small Container 5-ft. x 1-ft. 9-in.; 1 "Otto" Gas Engine, 6-in. bore, 14-in. Stroke; 1 Auxiliary Gas Compressor; 1 No. 5 "Crossley" Gas Holder; 1 400-light and 1 180-light Gas Meter, by "Cowan"; Galvanised Cisterns and Tanks; 1 12-H.P. Double Cylinder Gas Engine, by "Motorenfabrik Deutz"; 4 Sets Gearing; 4 Undercarriages for Cars; 1 Rochester Time Recorder; Oil Refiners; Chain Blocks; 48 10-Amp. 36-hours Oliver Lamps; 5 Line Resistances; 4 C.I. Boxes for same; 48 Arc Lamp Brackets; about 11 cwt. Copper Wire; Insulators; Straps; Switches; Benches; Vices; Drilling Machine; Grindstone and Frame; Wrought and Cast Iron Pulleys; W.I. Channel Girders; Brass, Iron and Lead Scrap; 2 High Pressure Coupling Pipes with Couplings; 150 Narrow Gauge Oak Trollies; Derrick Winch; several Tons Steel Rails, and other Effects; also the Wooden Erection forming

LARGE CAR SHED.
SALE TO COMMENCE AT 12 o'clock.
Without Reserve.

For further particulars apply to the AUCTIONEERS, 99, Whitworth Street, Manchester. Nat. Tel. 3990.

E. HULTON & CO., LTD., Printers, MANCHESTER.

The auction notice for the Trafford Park tram system in June 1908 included the four 12hp Deutz-engined tramcars, a horizontal engine by Tangye of Birmingham which drove the gas compressor, spare tramcar engines, four spare undercarriages for the tramcars, and all the equipment from the tramway's workshops. Even the wooden car shed was sold off together with 'several Tons Steel Rails' from the car shed and sidings.

Appointed to supervise the introduction of the gas-engined service was twenty-two-year-old William Henry Gaunt (1874–1951) who had served his apprenticeship with the Ashbury Railway Carriage & Iron Company of Openshaw who had built Blackpool's first gas trams.

The engineer placed in charge was Carl Lührig's onetime partner, Lucien Alphonse Legros, now Chief Engineer of the British Gas Traction Company, and he reportedly found the Trafford Park company very difficult to deal with. One of his more serious criticisms was that they seemed quite unwilling to consult with him about the specific requirements of running a gas tramway.

Taking into account the length of the wheelbase of the tramcars, Legros had advised that the maximum tolerable gradient on the line would be 1 in 33, and that no curves tighter than a fifty-foot radius could safely be negotiated. He had also specified that there would need to be a fuelling station – and a reliable gas supply from Salford's municipal gasworks – a workshop with inspection pit, and given the relatively light weight of the tramcars compared with a goods train, he required that the track be laid on very firm ballasted sleepers and a good solid weight of rail in order to ensure a smooth and stable ride.

If the tramway was not built to his exact specification, he had stressed, derailments would be inevitable.

What Holyoake found when he arrived was badly laid and poorly ballasted track on which the trams rocked dangerously, was shockingly poor quality workmanship, including straight rails where there should be curves and with loose or missing bolts where the fishplates joined the sleepers. With the scheduled opening just days away, he had also found that only one tramcar had been delivered, and because of the time it took to rectify those faults, the service did not resume until 8 April 1898, more than eight months after that first accident.

Indeed as late as December 1897 he had written to the estate company questioning whether or not the other three cars were ever going to be brought into service.

Presumably he was of the opinion that if they were going to be left languishing in the shed at Barton, he might consider moving them to Blackpool – the first of three such propositions between 1897 and 1901.

There would be a number of other problems during the lifetime of the tramway, largely caused by the shared use of the track by trams and freight trains, and a number of collisions were noted between the tramcars and Ship Canal Company trains, invariably caused by either trams or locomotives over-running the passing loops at along the shared sections of the three mile track. Trains, by virtue of their length, usually had right of way, but on a number of occasions locomotive drivers over-estimated the tramcars' ability to get out of the way.

By 1899 the British Gas Traction Company had gone into liquidation and Salford Corporation had cut off the supply of gas for the trams after non-payment of bills.

The service was once again suspended for a few months until the West Manchester Light Railway Company – which had been granted a Light Railway Order on 2 June the previous year – bought the equipment from the British Gas Traction Company for the knockdown price of £2,000 in 1900. It also took over the tracks and the Ship Canal Company's railway operations within the estate. The gas tram service continued under their management until it was withdrawn, with all the tramway plant and equipment being sold by auction in June 1908.

In the same year that the Trafford Park system opened, operation of the Neath Corporation Tramways route between Skewen and Briton Ferry in South Wales had also been sub-contracted to the British Gas Traction Company, with an initial fleet of four trams – and four more on order. They were also powered by Deutz engines.

In the absence of any records, it has been assumed that these cars were built by the Lancaster Carriage Works, and yet tramcar No.1 – the rebuilt vehicle which survives today – is one of the smaller cars which are believed to have been built by the Ashbury Railway Carriage and Iron Works. It can reasonably be assumed, therefore, that Car No.1 is one of the original Ashbury cars from the Lytham fleet acquired in 1903. But that poses a question – did Neath Tramways re-number their fleet when the

Augustus Krauss & Son's workforce replacing Neath's lightweight horse tram tracks with heavier duty rails to take the increased weight of the gas-engined cars. However, these look more like conventional railway metals than the standard profile of tramlines (see page 15). This picture is thought to date from 1898 or 1899.

Neath Tramways' Car No.1, built at the Ashbury Railway Carriage and Iron Company's works in 1897 or 1898, now displayed at Cefn Coed Museum, although the original plan was to make it the centrepiece of a new museum in the centre of Neath. As can be seen by comparing the two images, this page and opposite, the rebuilt tram does not quite replicate the end profile of the original vehicle – the end dashboards on the restored tram (right) come up to level with the bottom of the windows and go down to close to the roadway, whereas on contemporary photographs of the original tramcar in service (opposite page) the top of the dashboard is at least six inches lower, and the bottom would have been several inches higher to accommodate the required 'life preservers' – often referred to as 'cow-catchers' – front and back. This, and the increased height of the advertising panel around the stairway, unfortunately combine to make the restored tramcar appear much narrower than it actually is. For comparison, *see* page 127.

Neath Corporation Tramways Car No.1 on one of the passing loops on Neath Abbey Road – this one known was locally as 'Banana Island' and was near the site of the Roman fort of Nidum, parts of which have been excavated.

Blackpool cars arrived – for surely we might expect their original eight, acquired new in 1899–1900, to have carried the numbers 1 to 8. Unless, that is, the original gas trams bought from Lancaster Carriage were numbered in a sequence which followed on from the town's horse cars.

In that scenario, the numbers from former horse trams might have been reassigned in 1903 to ex-Blackpool stock.

As tramcar No.1 was still in service when the tramway was shut down in 1920 – by then having been in service for twenty-three years if it had come from Blackpool – claims of the gas cars being unreliable and difficult to maintain would seem to have been wide of the mark.

To handle the heavier weight of the new tramcars, in 1898 the entire track – which had previously been used by horse trams since around 1875 – was re-laid by Bristol-based engineers and contractors Augustus Krauss & Son, who had established local premises at 12 Lewis Road in Neath.

At around the time they would have completed re-laying the track, Krauss took out a classified advertisement in the *Evening Express*, advertising spare setts which they had removed from the roadway as additional passing places for the trams were installed.

Krauss are said to have been required – against their judgment – to re-use setts to save money on the project. That would subsequently become an issue.

The first test run of the Neath gas trams took place in late August 1899, under the watchful eye of Colonel Sir Francis Marindin, the Board of Trade Inspector and

Taken between 1910 and 1920, a rare photograph showing Neath Corporation Tramways Car No.1 being refuelled with gas outside the London Road tram depot. The numbering on this photograph of Car No.1 is quite different to both the view of the car on Neath Abbey Road, overleaf, and the rebuilt tram now at Cefn Coed.

'Senior Inspecting Officer of Railways', who was satisfied with what he saw. *The Neath Gazette* reported on 31 August that:

> 'On Wednesday afternoon a trial trip of the new gas motor cars which are to supersede the indifferent service which has been too long been endured, took place at Neath. The Neath Town Council can take credit as being the first municipal body which has

One of the large Neath Tramways cars turning into Orchard Street outside the entrance to Victoria Gardens. date unknown but probably just before the start of the First World War judging by the clothing.

BRITISH GAS TRACTION CO., LIMITED.

NEATH CORPORATION TRAMWAYS

THE public are respectfully informed that the Motor Cars will run on and from SATURDAY, the 2nd September, 1899, and

WILL ONLY STOP AT FIXED PLACES,

and intending Passengers are requested to get on and off at those places.

BRITON FERRY SECTION.

Square.   The Fountain (S. David's Church).   Tramway Depot.
Passing place, London road.   Windsor road.   Bowen street.
Exchange road.   School road.   Trick's corner.
Passing place, Penrhiewtyn.   Gas Lamp opposite 14, Penrhiewtyn
Lamp, at top of Penrhiewtyn hill.   Grandison Hotel.
Rhondda & Swansea Bay Station.   Ynysmardy Road.
School Passing place.   Grandison street, Briton Ferry.
Regent street (East), Briton Ferry.   Ritson street, ditto.

SKEWEN SECTION.

Neath Police Station.   Cribb's corner.   Top of Canal Bridge.
Top of Railway Bridge.   Dwryfelin Church.
Glynleiros Passing place.   Court Herbert.   Old Smiths' Arms.
Wesleyan Chapel, Neath Abbey.   S. John's Church.   Terminus.

BY ORDER.

WHITTINGTON, PRINTER AND STATIONER, POST OFFICE, WIND STREET, NEATH.

The handbill published by the British Gas Traction Company to inform 'intending passengers' of the new protocols which were being introduced with the gas tramcars. For some reason, between printing the handbill and the commencement of the service, the proposed tram stop on Ritson Street, Briton Ferry, was abandoned.

adopted the new system introduced by the British Gas Traction Co. Ltd. A splendid permanent way has been laid by Mr. Krauss of Bristol, and this and the car service were inspected and approved by an official of the Board of Trade earlier this week. So satisfied was he that it is understood that with the certificate to be forwarded in due course, there shall be a recommendation that the travelling speed of 8 miles per hour shall be conceded. This is more than has been granted to Blackpool or Trafford Park, where the service has been in use for some time past.'

That was a curious remark as there had recently been a dispute between the local authority and Krauss over the delayed introduction of the service – requiring the re-introduction of horse trams for a time – the former saying the track had been badly laid by Krauss, the latter, of course, saying it was the Corporation's fault.

According to the British Gas Traction Company, sub-contracted by Neath Corporation to run the trams, a further problem was that the Corporation had failed to provide a suitable gas supply. As at Trafford Park, another rather inauspicious start to running a gas tramway.

However, the 'Motor Car' service did eventually begin on Saturday 2 September 1899, and with it came an innovation for Neath – designated tram stops. The horse cars, which had run since 1875 could be hailed by waiting passengers and would stop on demand. The gas cars introduced a more formal timetabled regime.

The British Gas Traction Company was already in severe financial difficulty even before the Neath service started operating and it folded in October 1899, while its parent, The Gas Traction Company, continued to trade until dissolved in March 1911. The *Morning Post* newspaper reported on 2 December 1899 that:

'It will be remembered that the company, which recently went into liquidation, was formed in July 1896, to acquire the British patent rights of the Luhrig-Holt system of gas traction, the capital being £250,000. The chairman, Mr. G. S. Barnes, Senior Official Receiver, said that the accounts as finally settled had now been lodged. The total showed unsecured debts £39,610, mainly to the parent company. The assets, valued at £49,728, included the tramways at Blackpool, Neath and Trafford Park, and showed a surplus of £10,118 over the liabilities.'

A new operator, to be known as the Neath Gas Traction Company, took over the lease, but almost immediately changed its name to The Provincial Gas Traction Company. From the outset, the service it provided was beset with problems, so much so that the *Neath Gazette* was fanning the growing disquiet as early as February 1900:

'The recent eccentric movements, with many intermittent stoppages, of these motor freaks have called forth a great deal of indignation from a patient public who suffer from the delays en route, the lack of punctuality, the abandonment of stated service, etc. Of course there will be nothing done unless this matter of serious public inconvenience is brought into prominent notice, and it is doubtful, even then, if the Town Council will budge from the position of seeming indifference. But if nothing is done to remedy the present unsatisfactory state of things, matters will go from bad to worse, and what was intended to be for the public service will become a nuisance, and of no general utility.'

Given these comments, it must have seemed highly unlikely that the town would still be running a gas tram service up until 1920, having considerably added to its original fleet of four trams along the way.

When the Blackpool and Lytham gas trams were withdrawn, those of their cars which had survived the 1903 storm – possibly as many as three small and five or six of the larger vehicles, together with spares – were acquired by Neath and the fleet size thus increased from eight to at least fourteen and perhaps sixteen operational tramcars, but thereafter it entered a rather more stable period. A number of those ex-Lytham cars continued in use for twenty years, albeit reportedly in an increasingly disreputable state

Neath Tramways (ex-Blackpool, St. Anne's & Lytham Tramways) Car No.21, c.1900-1910. The advertisement for South Wales Furnishers on the headboard, has been recreated on the surviving gas tram in the Cefn Coed Colliery Museum.

There was an unusual division of responsibilities regarding the safe operation of the service. While the track and the streets along which it ran were the responsibility of the Corporation, the safe operation of the trams was the responsibility of the Provincial Gas Traction Company – a private company.

Amongst the hazards of the Briton Ferry to Skewen line were low bridges – not a good idea with a fleet consisting of open-topped double-decked tramcars. Needless to say, accidents occurred, as was reported in the *Cambria Daily Leader* on 18 July 1913:

'At the Glamorgan Assizes today, John Percival Smith, resident inspector of the Norwich Insurance Company, at Cardiff, brought an action against the Provincial Gas Traction Company at Neath for personal injuries while riding on the top of a tramcar between Skewen and Neath, on January 3rd, caused through the alleged negligence of the defendants. The defendants denied that either they or their

154 • THE GAS TRAMCAR

Left: The crew pose on Car No.19 as it stands at the terminus at Skewen. Car No.19 is showing its age in this photograph which was possibly taken not long before the system was shut down. The bodywork is sagging quite severely either side of the bogies. Despite the number of tramcars owned by the Corporation, it seems only to have had a maximum of nine drivers and a similar number of conductors on the payroll.

Opposite above: Neath's larger fifty-two-seater Car No.20, with a smaller forty-seater beyond it in the tramshed in 1920, just after the service had been withdrawn. At this point, the entire fleet and all the spare parts were about to be sold by auction. This suggests that, at closure, there were still at least two of the smaller Ashbury cars in the fleet. Most of the fleet went for scrap and were broken up, all except that one car which was rescued from a private garden in the 1980s. The small car carries the legend 'Neath Corporation Tramway' on the side panel, whereas on the rest of the fleet – and the restored Car No.1 – it reads 'Neath Corporation Tramways'.

Opposite below: One of the spare Otto-cycle tramcar engines built for the British Gas Traction Company, built in accordance with Henry Holt's modified specification by Gasmotoren-Fabrik Deutz, lying in the Neath depot around the time the service was discontinued in 1920. To the left is one of the metal cylindrical tanks which contained and protected the collapsible gas bags.

servants were guilty of negligence, that that the plaintiff himself was guilty of contributory negligence…Mr. Llewellyn Williams said the plaintiff travelled from Skewen to Neath on the outside of the car. When he came to Bridge-street, near the Great Western railway bridge, the plaintiff, who was sitting on the front part of the car saw the office of Mr. Bevan, with whom he wanted to do business, and at the same time he saw a notice on the parapet of the canal bridge stating that the car would stop there if required. He got up from his seat, and as there was no means on the top of the car to give the signal that he required to get off, he proceeded to make his way to the back of the car. He walked along, his back being towards the railway bridge, and after he had taken about three steps he was struck on the head and fell on his face on the top of the car. He had received a severe blow at the back of the skull which made him insensible. When he came to himself, he crawled on to a seat and when the car stopped, he made his way downstairs. Having alighted in a dazed condition, he went to the Tramway Offices and saw Mr. Imray, the sub-manager, and told him that he had been struck by the bridge and had not been given warning. Mr. Imray replied that he must have been given warning, and he further stated that they were not responsible for giving passengers warning, and that the plaintiff was lucky to have got off so lightly, as others who had been struck had been cut and covered with blood…'

This was clearly a time before a tramway company felt any duty of care towards its customers. As the company under its various ownerships had been operating the route with double-decked cars for at least sixteen years by then – and Imray's response implied that such an accident was not a unique occurrence – one might have expected the management of the Provincial Gas Traction Company to have rather responded rather more positively, putting systems in place to ensure it did not happen again.

But it would appear that this was by no means their first accident – the *South Wales Echo* had reported a serious incident in its issue of 20 November 1900, just over a

Right: Car No.3 at the line's Skewen terminus – from a rare chromolithographed postcard probably published between 1905 and 1912. The line was single track with passing loops – twelve in all, including one at either end of the line. The publishers of this expensively produced card must have believed they were going to sell it in sufficiently large numbers to warrant the additional expense over a sepia card. Postcards of gas trams in action are scarce enough, but this coloured one is believed to be unique.

Below right: A model of one of the larger trams, made in 1995 by Harry Barnsley, Model Maker, and presented to the Mayor of Neath on behalf of the Neath Borough Council Training Agency. Mr. Barnsley chose a brown and cream livery for his model – much closer to the maroon and cream paintwork found on the original tramcar than the green and cream it now carries. Barnsley's model accurately reflects the height and shape of the dashboards, but without the 'life preservers' or 'cow-catchers'. The model is displayed in Neath Library.

year after the service was inaugurated. Headlined 'Motor Cars in Collision', the report read:

'A very ugly accident occurred on Monday near the Smiths' Arms, Neath Abbey. One of the gas motor cars working on the Neath and District Tramways ran wild on the Skewen Hill. The speed of its descent was terrific, and near the foot of the hill, it nearly ran down a horse and cart, which was barely pulled out of the way in time. A representative of the "South Wales Daily News" was in the runaway car, and with other occupants heard the driver shout to the conductor to apply the second brake. This the conductor failed to do, and the car darted through the village of Neath Abbey. Unfortunately a laden dray belonging to Mr. E. Evans Bevan, of the Cadoxton Breweries, was crossing the tramline, and into this dashed the motor tramcar. The horse was knocked down, and such was the force of the impact that the barrels of beer were jerked into the air. The drayman was extricated from the wreckage of the dray. He was in a sad condition, his right leg being broken, and his head seriously injured. The tramcar was much damaged, all the glass on one side being shivered and the front brake, so damaged, that it could not work. The drayman, whose name is John Hughes, was with all speed attended to by Dr. Whittington, and shortly afterwards was conveyed to his home.'

It is worth noting two things about that report – the first of which was the lack of hospitalisation for the seriously injured drayman. Secondly the fact that a runaway car had been able to descend the hill at speed with such a serious effect, suggests quite a steep gradient which it would previously have had to run up as well – a fact which contrasts sharply with the often-repeated claims that the gas trams could not cope with even gentle slopes. Such claims, presenting 'opinions' as 'facts', were then sometimes fanned – as they still would be today – by those who had a vested interest in promoting alternative traction systems, especially electricity.

The locals continued to use the tramway, accepting its shortcomings – and quite high fares at a penny a mile – with good humour, most of the time. There was even an anonymous piece of doggerel composed, summing up the experience – and echoing similar passenger experiences at Maastricht and elsewhere:

'Oh! the cars, the Neath tramcars
When they are in motion,
You think you're on the ocean.
The seats are very handy for people up above
But when you get to Skewen Hill
You've got to get out and shove.'

There were the obvious drawbacks to the gas tramcar – limited range, exhaust fumes and so on, but compared with what the world would go on to endure from petrol and diesel vehicles, the exhaust fumes were hardly an issue at all – just a change from the smell of horses and steam wagons. But opponents of the gas cars amplified the issue in support of their position.

Top: A 1924 7hp Austin Seven 'Chummy'. At just 48 inches wide, this car would have slipped easily inside the tramcar with more than 10 inches clearance either side.

Above: A 1969 model Hillman Imp fitted with the 'de luxe' wing mirrors measured a little over 61 inches wide, so would have been a very tight fit.

By 1916, with the fleet approaching twenty years old, the tramway was operating a four-car service at busier times, with just three cars during off-peak periods, and to maintain reliability, some of the many cars in the shed were being pirated for spares to keep the others running.

By 1920, with the trams' mechanical equipment worn out, the entire tram system was shut down, to be replaced by motor buses.

And there the story of the Neath's gas tramcar's might have ended and been forgotten, as all the cars and ancillary equipment were sold as scrap. One of the small ex-Lytham, St. Anne's & Blackpool tramcars, however, survived.

At or after the disposal sale, Car No.1 was acquired by local motor car and motor cycle dealer Matt Price who had premises on Main Road, Neath Abbey. Just how long he kept it is unclear, but at some point he sold the body shell to Frank Beddoes, a former tramway company mechanic, who planned to use it as a garden workshop and garage.

Mr. Beddoes paid just one pound for it, and had it moved to his home. What happened to the underframe, wheels, suspension, engine and other mechanicals has not been recorded, leaving no surviving example of the gas cars' powertrain. Nor has a date been found for exactly when he acquired it.

Newspaper reports at the time of the tram's rescue offer a variety of dates between the 1920s and the 1960s for its conversion.

By that simple piece of serendipity, Car No.1 survived, although not in any form which immediately said 'tramcar', and there it sat until rediscovered in 1982. A casual passer-by could have been excused for not recognising what it had once been.

To convert it into a garage, all the interior fittings had been removed and the upper deck had been discarded, one end had been completely cut away, and one of the end platforms had been removed.

A large door had been cut in the engine side of the body to enable Mr. Beddoes to get in and out of his car – even a small car would have barely fitted within the narrow tram body. That side entrance was protected from the elements by the addition of a lean-to. A pitched roof had been added, and the whole vehicle liberally tarred. Another lean-to added at one end became a workshop.

The photographs taken at the time of its rescue in the 1980s show just what a leap of faith was being made by those who believed it could be restored.

Today, undertaking such a conservation project would be handled in a very different manner – at much greater expense and using a very different approach to the preliminary research which was undertaken before the rebuild started.

It reportedly took two years of secret negotiations before the wooden shell was craned out of the garden and transported to the Council's Training Agency workshops to begin the long process of reconstruction. It was apparently well bedded into the ground, with considerable rotting of the lower timbers.

Opinions differ as to what sort of car Mr. Beddoes might have planned to keep in it – and without knowing the date of the conversion it is impossible in determine his choice. A retired tram mechanic would probably not have bought a new car and, depending on when he bought his first car, there would have been a choice of small models from which to chose.

There were a limited number of cars on the market during the time period in question which would have fitted inside. With the seats and the flywheel housing removed, the interior width of the vehicle was just over 70 inches.

If it was the late 1920s or the 1930s, the most popular small car of the time – with over 300,000 of them built – would have been the Austin Seven. The width of the earliest 1922 model was 48 inches, giving a clearance of 10 or 11 inches either side.

Interestingly, small cars such as the Austin A30 and A35 (1951–68), A40 Farina (1961–67) the original Mini (1959–1999) and the Metro (1980–90) would all have fitted inside, so the tram could have continued to house cars well in to the 1980s.

The rescued shell of the tramcar being rolled into the Youth Training Scheme workshops on Milland Road in Neath in early December 1984. There it was rebuilt by a group of apprentices and later emerged as the vehicle which is displayed at Cefn Coed today. For the previous sixty years it had been used as a garage in the Melincryddan area of the town by a retired tram driver, who had bought it for just one pound to house his car – some sources say in 1920, others 1924. Part of the rescue deal was that in return for handing over the tram's body shell, Neath Borough Council paid the owner at the time the cost of a brand new garage for her car.

According to one report, the last of Mr. Beddoes' cars to be garaged in the tramcar was a Hillman Imp Mark 1 and that would have been a remarkably tight squeeze as, at fractionally over 5 feet wide, it would have had just 4 or 5 inches clearance either side of the tram's side walls. The original Imp had no wing mirrors – fitting them added another two inches, making the fit even tighter.

Amongst the surviving letters and documents from the time, there are some which reveal some interesting facts – and some oddities. For example, on the surviving flywheel door, traces of maroon and cream paintwork were discovered, leading some of the team to suggest that the tram must originally have carried that livery.

Local newspaper stories at the time the tram was restored tended to contradict each other, with one man reported in the *Neath Guardian* in 1987 saying the original fleet was painted 'bottle green and cream', while a former conductor complained to the *Western Mail* that the Neath trams he worked on were 'cream and chocolate brown'. So perhaps the maroon paint was in fact faded brown, and Harry Barnsley's model – *page 157* – actually reflects the original Neath livery.

However, why the rebuilt tramcar finished up being painted green and cream – a colour scheme it would never have carried under either Lytham or Neath ownership – remains a mystery. The livery looks much more like that which would have been carried by Crosville buses of the period.

The letter which reported the 'maroon' paint also remarked on 'the meticulous attention to detail, thus ensuring that every little part of the tram is restored precisely correctly' – a statement at odds with the acceptance a few paragraphs later that it had been impossible to source wheels of the correct size, the recycled wheels from a Great Western Railway guard's van acquired from Barry scrapyard being 'too large and heavy'.

It is also clear that, for whatever reason, a number of quite significant errors were made in the new platform ends and dashboards of the vehicle.

Luckily, the beautiful vaulted wooden ceiling of the lower deck, with its steamed and bent wood had largely survived, and equally fortunately, numerous photographs survived to give an idea of what the finished vehicle *should* look like.

Some other compromises to the design were introduced, such as raising the height of the shielding to the stairs which carried the advertising banners – perhaps reflecting the beginnings of a health and safety consciousness which had been notoriously absent sixty years earlier – as reported in the story on page 152.

Such safety features would have been considered essential even back then, especially as small children would be likely to explore the restored tram's upper deck.

It was accepted from the outset that the acquisition of a gas engine and transmission was 'right out of the question', that being a major engineering undertaking.

Perhaps the restorers' greatest disappointment would have been their inability to realise one of the two key objectives of the rescue. The first had been the restoration of the vehicle as a static exhibit, but the second had been to 'install it as the centrepiece of a small new museum to be built in the centre of Neath town'.

That would have allowed it to be presented and explained in a much more complete manner. As it is, the preserved vehicle does not really reflect what a gas tramcar would actually have looked like back in the early twentieth century.

It would be good to think that at some time in the future, funding might be found to correct the 1980s errors, and that a Trust might be set up to fundraise and realise the idea of presenting the tramcar in its own town-centre museum as originally planned.

But seventy years before the rescue was even contemplated, Neath was not the only town where gas-powered tramcars worked. While the Neath Tramway had made a conscious decision from the outset to use gas as a fuel, elsewhere that decision was simple expediency. In Morecambe in Lancashire, the Morecambe Tramways Company Ltd adopted a gas-fuelled system as the only way of keeping the service operational in a time of fuel shortages on its short and more rural line between the Battery Hotel and Strawberry Gardens – a route which actually lay entirely within neighbouring Heysham.

The company had retained the one-and-a-quarter mile line when it sold most of its network to Morecambe Corporation, and up until 1912, it had been served by horse trams. But as the town's popularity as a tourist destination grew – and the number of pedestrians increased – the challenge of keeping the streets clean was also becoming an issue.

The directors decided to modernise the service, while making a trip on the tramway a tourist attraction in its own right – not riding a tramcar pulled by horses or steam engines, nor powered from overhead wires, but riding on a self-contained tramcar, the first in Britain to be powered by a petrol-burning internal combustion engine.

They commissioned four vehicles from the United Electric Car Company of Preston – three enclosed single-deckers, and an open 'toast-rack' car. All were fitted with the Leyland four-cylinder 55 horsepower engine – later reduced to 44 horsepower – introduced in 1910 and fitted to Leyland's first fire engine.

The transmission was modified for the trams so that all four gears were available with the car travelling in either direction. With a driving position at either end of the vehicles, a locking pin was used to isolate whichever set of controls was not in use.

The four cars were not the world's first petrol-fuelled trams – that accolade goes to David Brown who had built similar vehicles for the Karachi tramways in 1911 – but the story of how the Morecambe trams were modified so they could continue in use during the First World War as supplies of petrol became increasingly limited, earns them a place in this book.

By the summer of 1917, petrol was in very short supply, and a number of alternatives were being considered. *The Western Mail*, on 8 June 1917, drew its readers' attention to the obvious alternative:

> 'Coal-gas as a substitute for petrol has for some time been engaging the attention of the motor industry, and a number of successful experiments have been made. Neath Tramways, for example, have been running on coal-gas for eighteen years, using a storage pressure of 200lb. per square inch.'

In early 1918, the Morecambe Tramways Company converted their three enclosed cars into dual-fuel vehicles, able to run either on petrol or town gas, leaving only the open 'toast-rack' summer car exclusively burning petrol, and according to an account of the conversion in *The Commercial Motor* on 29 August 1918, the driver on the dual-fuel cars could quickly switch back to petrol if the gas supply ran out, but as the tramway had funded the conversion to town gas because it could not rely on getting access to any petrol, that capability would have been of little consequence.

The conversion was straightforward – burning gas rather than petrol, there was no need for the carburettor and that was bypassed with a simple regulator installed alongside it to control gas supply.

162 • THE GAS TRAMCAR

One of several postcards of the petrol-burning internal combustion engined tramcars – advertised as 'the first in the Kingdom' – built for the Morecambe Tramways Company Ltd by the United Electric Car Company, which became a subsidiary of Dick, Kerr & Company in 1917. According to *The Commercial Motor* magazine, they cost under 8d a mile to run with up to sixty passengers.

Gas was stored in 900 cubic feet rubber container on the roof of each car, protected by a wooden framework which contained the collapsing tank and avoided it flapping about as the fuel was used up. With no low bridges on the route, the increased height of the tram was not an issue.

Eight round trips could be completed on a single charge of gas, supplied via a flexible hose at the depot. The engine's power was reduced to 40hp when it was running on gas, but that was still more than powerful enough when operating such a light vehicle on that route.

The service was inaugurated in March 1918, but a report in *The Visiter* newspaper on 25 September told of a sign on the company's offices announcing 'No car – no petrol' and another – the day before *The Commercial Motor* article – said two cars were out of service due to shortage of spares.

Petrol remained in short supply for several years after the end of the war, so it is possible that the tramcars continued to run on gas at least until 1922. As the cars were withdrawn in late October 1924 and replaced by electric trams, perhaps using gas remained an option until the end.

None of Morecambe's trio of innovative gas trams has survived, nor has the open toast-rack car which remained petrol-fuelled. Despite claims to the contrary, they were the last gas-fuelled tramcars to run anywhere in the world for almost one hundred years.

Several sources have suggested that gas trams were introduced on to the street of Tallinn, Estonia, in 1921 but this is not borne out by evidence, and the claim may arise from a confusion in translation between 'gas' as understood in Britain and 'gas', meaning gasoline, as understood in the United States and elsewhere.

Now, just over a century after the last gas-engined tramcar ran on a British tramway, is the story about to go full circle?

THE COMMERCIAL MOTOR — 29th August, 1918

furnace is first lit, there is a certain amount of smoke produced, it is necessary to provide some means for getting rid of this. The inventors have incorporated a small cock in the gas pipe, which is placed near to the induction manifold, and in the beginning, when the fire is first lit, this cock may be left open. When the engine is actually running, this tap may still remain open to a greater or less extent, providing an explosive mixture for the gas drawn from the producer. Provision for cleaning out the producer is made, an air-tight door giving access to the interior of the firebox, while a small valve allows of dust being removed from its interior.

A TRAMWAY SERVICE ON GAS.
Morecambe Tramway System Makes a Success of a Flexible Container.

The war has brought into prominence in many parts of the world the value of rail or tramway cars driven by internal-combustion engines of the motor lorry type. True, most of these war-time innovations, the majority in the hands of the military authorities, of course, have been adapted from any kind of spare parts which came to hand, and were fitted with an engine built for an entirely different purpose from that to which it was being put, and often with a make-shift transmission.

Yet, even in these adverse circumstances, this form of transport has proved its worth, particularly in a country ill-supplied with roads, and where either rails were already laid or the laying of them was a more simple matter than the elaborate preparations which have to be made for the construction of a road capable of withstanding the stresses of heavy and fast-moving traffic.

In the great development of the sources of raw materials necessary for the process of reconstruction after the war and the extension of colonial enterprise which is expected, there will, doubtless, be an opportunity provided for the British makers who are prepared to take up this question and to produce a vehicle suitable for the need, having power unit, transmission and all working parts proportioned and balanced for work under conditions which are not identical with any other form of service.

Already, before the war, one or two English concerns had seen the possibilities looming ahead, and although most of the work in this direction has been for export, there is, running in this country, a short length of tramway worked on this system. The Morecambe Tramways Co., Ltd., a private concern, with powers to carry passengers over a length of line 1¼ miles long in the Heysham District Council's area, decided in 1912 to transfer from horse vehicles to power, and the suggestions of Leyland Motors, Ltd., that a petrol service be provided were adopted. The Leyland company, with their

Morecambe gas-driven trams.
(1) Showing one of the complete 55 h.p. Leyland petrol tramcars with 900 c. ft. flexible gas-bag fully inflated. (2) Filling up outside the depot in order to minimize the risk of fire. (3) The gas supply from container to engine is taken off the filler pipe: a flexible metallic pipe runs along the canopy and down the front of car to the engine. (4) Front view of car, which retains a symmetrical bearing.

Several publishers produced black-bordered 'remembrance cards', lamenting the replacement of horses or steam engines with all the overhead wiring involved with electric cars. Northampton had run horse trams from 1880 until 1904 before electricity was introduced on to the network.

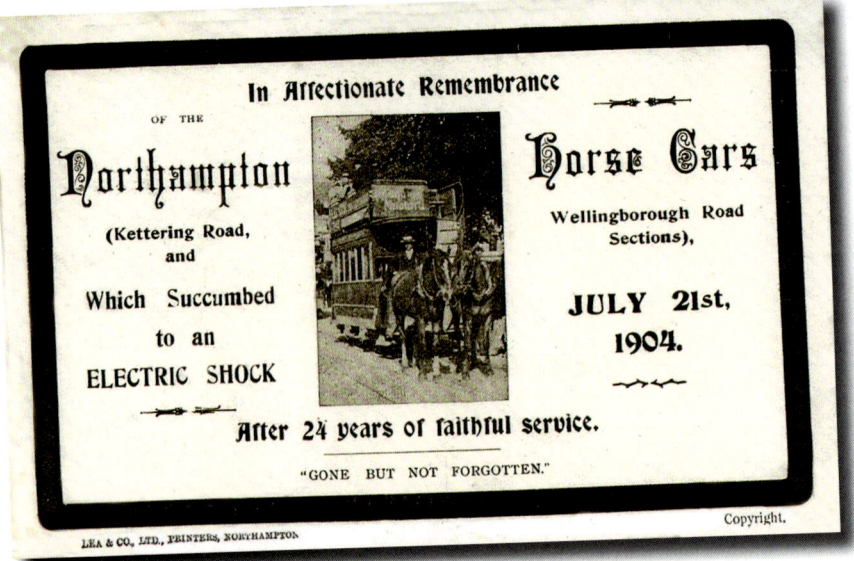

The Derby Horse Cars had also operated for around a quarter of a century before the system was electrified in 1904.

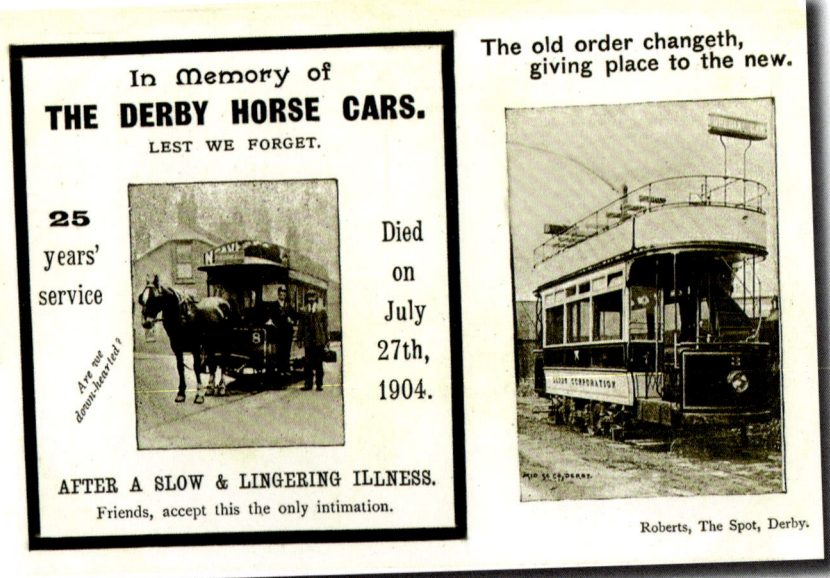

Birmingham's Kitson steam tram locomotives were replaced in 1907 – again not to everyone's approval. Their trailer cars were built by the Falcon Car Works in Loughborough. The same picture was used to commemorate the end of steam on the Saltley, Perry Bar, Witton and Lozells route – *see page 35*.

ELECTRICITY WINS THE DAY

Back in the early 1880s, at the same time that Henry Holt, Frank Crossley and others were exploring the practicality of fitting gas engines into tramcars, a completely different approach to exploiting the power of such engines was being developed by Magnus Volk. Instead of using the gas engine as the primary power source on board his tramcars, he proposed using it to generate electricity.

When the Tramways Act had passed into law in 1870, electricity as a reliable power source which could be available twenty-four hours a day was still a thing of the future. Even in the closing decades of the nineteenth century, only a very few towns and cities had any sort of stable electricity supply, let alone one capable of generating and distributing sufficient capacity to support electric trams. Thus horses and steam engines were the most efficient options available.

When the time came to replace horses, Sheffield Corporation and the authorities in Paris stand out as having taken the most logical and structured approach to determining which form of traction best suited their specific needs. Those investigations would not take place until the 1890s. Until that time, the number of horses pulling tramcars and omnibuses increased dramatically as towns and cities introduced and extended their tramway networks.

Given the fact that small internal combustion engines had limited power output before the First World War – after which time it was the development of more powerful petrol engines that took precedence – gas-engined tramcars really remained no more than a very good idea which was still waiting for technology to catch up, despite their many advocates in the 1870s and '80s. But by the time that technology did catch up, and compact powerful gas engines became available, the investment in electrical traction had progressed beyond the point where other systems might be given a chance.

As the nineteenth century drew to a close, however, what might have been available on the early days was an intermittent DC supply, and so horse cars, accompanied from the 1880s by steam tramways, remained in widespread use.

In Britain, Brighton was the first to install electric street lighting, on 10 February 1881, when it was introduced experimentally by Charles Siemens who erected a series of electric-arc lamps along Marine Parade, with a generator installed in an archway below. The dynamo was powered by a steam engine. Later that same year, a concert at the Royal Pavilion became the first such event to be lit by electricity, using a dynamo driven by a converted steam road roller, under the direct supervision of Magnus Volk – who had also been involved in the Promenade lighting experiment and would later become a very well-known figure in Brighton's tourist industry.

Brighton's lighting experiments were followed by the opening of the Edison Electric Light Station in Holborn,

While the majority of the commemorative tram postcards were of the 'in Memoriam' style, lamenting the passing of either horse trams or steamers, some postcards celebrated the launch of electric cars in 1907. The passing of the gas car services in Blackpool, St. Annes and Lytham, and later in Neath, appears to have passed without the publication of any souvenirs.

166 • THE GAS TRAMCAR

Above: One of many Edwardian postcards showing Volk's Electric Railway, the first electric tramway in Britain, which had been opened in 1883. Electricity was originally provided by a 2hp Otto gas engine driving a Siemens D5 50 volt DC generator, giving the tramcar a top speed of 6mph.

Right: The Black Rock Terminus with two of the original small tramcars on station. Today, Black Rock has a much more impressive terminus building.

London, the following year, its electricity being generated by a coal-fired steam turbine that produced enough power for around a thousand street lamps. The project consistently lost money and closed in 1886. For the next few years, Holborn and Holborn Viaduct reverted to being lit by gas.

The first electric tramway was trialled in Saint Petersburg in 1875, with Siemens demonstrating a practical system in Berlin in 1879.

Magnus Volk's Electric Railway opened in Brighton in 1883. It is credited with being the first in Britain and, still operating today, is now celebrated as the oldest working electric tramway anywhere in the world.

From the point of view of this book, its interest obviously lies in the fact that for the first few years, before Brighton had a robust enough power supply, the electricity the tramway used was generated on site in another of the archways beneath the Promenade. In this case, however, it was not a steam engine which drove the dynamo but small Otto-cycle gas engines connected to Siemens direct current generator sets.

The first 2 horsepower engine – installed in 1883 – was replaced by an 8 horsepower engine in 1884 when the track gauge was widened from 2 feet to 2 feet 8½ inches, and the original Siemens D5 50 volt DC generator was replaced by a Siemens D2 40 amp 160 volt set. Before 1888 that had in turn been replaced by a 12 horsepower engine – more than sufficient to run two cars along the 1,400-yard line. It is assumed that the 12 horsepower engine and generator also delivered 160 volts.

Power was initially delivered direct to one of the rails, the other being used as the return, but after less than three years that was replaced by a live third rail in 1886, with both the track rails acting as returns. A detailed description of the system was carried in Crossley's 1888 catalogue, which reported:

'Some little time ago we were favoured by Mr. Magnus Volk with the following brief but interesting particulars reflecting the above railway. The railway, which is a quarter of a mile in length, and having a grade (going out) of 1 in a100, commenced

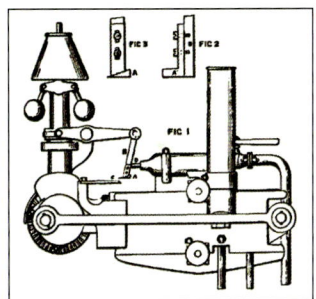

Top: Opening day for Volk's railway, 3 August 1883, with Magnus Volk himself standing on the tram's left-hand platform.

Above: Magnus Volk adapted a small centrifugal governor to control the speed and consistency of the Crossley Brothers gas engine driving the Siemens generating set that produced electricity for his railway.

Left: The compact 2 horsepower Crossley engine with its generator set attached. Sited in one of the arches below the Promenade, this design of engine was used to generate electricity for Volk's original ¼-mile long electric tramway. His modified governor can be seen at the left end of the engine. It was replaced by a larger Crossley engine when the line was extended in 1884.

168 • THE GAS TRAMCAR

The tramway companies in Sheffield (right) and Ipswich (middle) had both considered gas traction before deciding on electricity, introducing overhead electric systems in 1899 and 1903 respectively.

High Street, Sheffield.

ELECTRIC TRAMS IN IPSWICH. A TRIAL TRIP, NOVEMBER, 1903.

Preston introduced horse trams in 1879 and their first electric tramcar ran on 7 June 1904 – an event celebrated on local postcards. The power station on Holmrook Road, close to the Deepdale Road Tram Depot, housed generating equipment by Dick, Kerr & Company of Preston and Kilmarnock, who also built the cars and laid the track.

Preston's First Tram.

work on August 4th 1883. From this period up to September 29th 1883, there were in all 24,000 persons conveyed in the cars—about 10 or 12 passengers a time. The power is obtained from an "Otto" Engine of 2-HP (nominal), driving one of Siemens' D 5 Machines. The gas bill during the above-mentioned period, i.e., August 4th to September 29th 1883, in running the Engine (working 10 hours per day, and making six journeys per hour) amounted to £4.12s., including that used for the side lights, and one burner used to light the arch; cost of gas, 3s.3d. per thousand cubic feet. Mr. Volk expresses himself highly satisfied with the engine.

Since the above was first published an Engine of greater power, i.e., 12-HP nominal, has been substituted, and the line extended to one mile in length. This Engine drives two cars seating thirty persons each; cost of gas at 3/3 per 1,000 cubic feet, 11d per hour.'

The more powerful engine and a larger Siemens generator set continued in use until the system was converted to mains electricity.

Blackpool's first electric line opened in 1885, picking up its electric current from a conduit between the rails. The conduit channel suffered from regularly being filled up with sea water and sand, causing power leakages and even short circuits and it was abandoned by 1899, being replaced by overhead wires.

Electrical pick-up from overhead cables was first pioneered on a short line near Vienna in Austria in 1883, a system which, in a more reliable form, would become standard across much of the world.

The introduction of electric trams may have had its vocal and enthusiastic supporters, but for the residents of many towns and cities the infrastructure of the new system was considered an eyesore, with some councils banning the use of overhead cables.

The introduction of Accrington's electric tramcars was celebrated in postcards – as had been their steam trams introduced by the Accrington Corporation Steam Tramways Company as early as 1886 (*see page 28*). The first batch of eighteen Brush electric cars was introduced – and Car No.6 photographed for this postcard – in 1907. The tramway operated until 1932.

As has been discussed earlier in this work, for the tramway companies, installing an electric system to replace either horses or steam locomotives was expensive and disruptive. There were even those who openly stated their preference for retaining the steam trams, had it not been for the dirt and smells associated with them.

While horse traction was difficult to defend in terms of pollution and filthy streets, steam was clearly not ideally suited to the narrow streets along which tramways were being operated, and with compressed air, gas or oil perhaps it was just that these were years ahead of their time. Cables and conduit systems were expensive to maintain and potentially dangerous.

Parts of London refused to allow overhead catenary, while permitting the potentially rather more dangerous conduit system with electrical pickup below the street surface. New conduit lines were still being installed in London in preparation for the 1951 Festival of Britain, but the last such line closed in 1952.

Ultimately, overhead wires won the day across most of Britain and despite being based on technology introduced a century and a quarter ago, it is still the preferred system for the current generation of tram and light rail systems.

Postcard companies celebrated the arrival of the new electric trams with great enthusiasm,

Above: Restored Glasgow Corporation tram No.1017 was built in 1904 for Paisley District Tramway as an open top double-decker. In 1923 Glasgow took over the Paisley services and 1017 was converted into a single decker. The power collector tower dates from 1925, after which it was generally used as a driver-training vehicle. After a ten-year restoration by the Summerlee Transport Group, the tram returned to service in 2002 at the Museum of Scottish Industrial Life.

Right: The restored ex-Sunderland tram No.16 is one of a small fleet which now takes visitors around the vast Beamish Museum site in Northumberland.

Left: Rochdale in Lancashire inaugurated its electric tramway on 22 May 1902, replacing earlier steam services. The system was eventually extended to Bacup, Castleton, Newhey, Norden and Todmorden, but was closed down in 1932.

Below: Wolverhampton Car No.49 was built in 1909 by the United Electric Car Company of Preston, originally with an enclosed upper deck, but now open-topped in in preservation. It can regularly be seen running on the tramway at the Black Country Living Museum in Dudley.

producing tinted cards in their hundreds – the mass adoption of electric trams and the adoption of the postcard as the universal means of speedy communication in Edwardian Britain happened at around the same time.

Neath, Trafford Park and Blackpool were still operating gas trams when the postcard boom got underway, the Blackpool, St. Anne's & Lytham Tramways Company replacing them with electric cars in 1903, Trafford Park in 1908 and Neath closing its tramway completely in 1920.

However, despite postcard companies being keen to mark the demise of steam trams and horse trams, postcards of those three tramways in operation were few and far between and, unlike the passing of horses or steamers, there appear to have been no cards published to mark their passing.

One Lytham card was published as part of the 'Reliable Series' (*see page 135*) and the single known postcard view of a Neath tramcar at Skewen was produced in colour (*pages 156–7*) and publishers did not go to the expense of chromo-lithographic printing if they did not anticipate healthy sales.

After 1920 when the Neath fleet was withdrawn, the gas-motor tram, an idea more than a century ahead of its time, was consigned to the history books and largely forgotten. During the century which followed, the expanding tramway network was entirely electric.

Now world transport is being directed towards more and more – and heavier and heavier – vehicles running on rubber tyres which give off potentially much more damaging particulates than carbon dioxide as they wear out their treads, so trams are, per se, cleaner and more environmentally friendly however they are fuelled.

But is the time right for the return of the gas tramcar, albeit using twenty-first century technology?

CAN THERE BE A FUTURE FOR GAS-ENGINED TRAMCARS?

Back in 2014, passengers arriving at Bristol Airport bound for Bath boarded an unusually decorated bus, the decoration chosen by GENeco and Wessex Water to promote their 'green' bio-fuel. Along one side was an illustration of people sitting on toilets. It was certainly eye-catching. The fuel was biomethane, derived solely from human and food waste. This was the latest in a growing number of gas-powered vehicles across the world, reflecting the fact that bio-gas has an important role to play in improving air quality.

Research into 'greening' transport has been going on across the world for more than twenty-five years, and Scania's first 'green' bio-bus was launched in the 1990s. Bus fleets in Bristol, Reading, Nottingham and elsewhere have all converted to biomethane, benefitting from the government's 'Low emission bus scheme'. If it works for buses – and also for many of the other vehicles in the transport industry – then why not for trams? Trams running on steel rails are, by the nature of the surfaces on which they operate, much more efficient than buses due to the flexing of bus tyres on tarmac. The rolling resistance of tram wheels on rails is 85 per cent less than for buses – and without the particulates released into the air by both the granulation of both the tyre itself and degradation of the road surface.

A century after the demise of the first-generation gas tramcar, viable alternative power sources to the petrol or diesel engine – and indeed to the conventional electric tramcar with its overhead catenary – are being widely trialled around the world.

Innovative technology, now moving beyond the experimental stage, offers the prospect of a viable gas-fuelled hybrid system, and the trams, buses and ultra-light railcars on which the system is being tested are yielding encouraging results.

Opposite: A gas tram of the future? Ultra Light Rail Partners Ltd are proposing two gas-powered systems. They envisage a three-car tram unit powered by a hybrid system where two small 0.9 litre biomethane-fuelled engines will power on-board electric generator sets charging batteries which, in turn, will supply the electric motors in each of the vehicle's bogies.

Left: The Scania biomethane-fuelled bus which was promoted by Wessex Water and GENeco in 2014 was one of a number of initiatives across Europe which demonstrated that an eco-friendly gas-powered transport system was both possible and practicable. Biomethane is fed into the national gas grid, and the bus operators can draw exactly the same quantity out – a mix of natural gas and biomethane – and still call it bio-fuel.

Ultra Light Rail Partners Ltd is developing a hybrid tram system where small on-board biomethane-fuelled engines drive electricity generating sets to charge the batteries which, in turn, provide power for the vehicle's electric driving motors.

One proposed design for their articulated tramcar would carry its gas tanks in the roof – interestingly the same layout which Carl Lührig proposed in his early patents more than a century and a quarter ago.

But whereas the 'town gas' tramcars which Lührig developed had a very short range between fuelling stops, modern technology makes it possible for the gas to be safely compressed to a much higher pressure than was possible a century and more ago. While compressing 'town gas' to about eight times atmospheric pressure was the maximum possible in the 1890s, modern cylinders can contain biomethane at over 200 times atmospheric pressure. That means that the vehicle's range will be much greater than its Victorian and Edwardian predecessors – up to 1,000 kilometres is predicted for the ULRP vehicle.

By comparison, the purely battery electric variant of the proposed design would have to recharge its batteries at rapid charging points much more frequently.

In the biomethane light rail car which ULRP is also proposing, the higher ride height of railway vehicles would enable the gas tanks to be carried beneath the floors – as was the case with both Lührig's and British Gas Traction's production tramcars back at the start of the twentieth century.

In the bio-gas version, the gas motors, generating sets, and the batteries would sit beneath the driving positions forward of the bogies at either end of the vehicle. In the electric variant, all of those spaces would be taken up by the batteries.

Below: Christopher Maltin, of Biomethane Ltd, fuelling the ULRP test vehicle on its first successful run at Long Marston on 22 July 2020.

Bottom: The test vehicle is a modified Class 139 railcar from Parry People Movers, with its 2.4 litre Ford propane engine converted to use biomethane. The original PPM hybrid vehicle has been running between Stourbridge Junction and Stourbridge Town since 2009.

Below right: The biomethane engine and drive train on the test vehicle.

The biomethane tram will harvest energy from the vehicle's braking system to further charge the on-board batteries and thus increase the maximum range before the gas tanks need replenishment.

So, what is 'biomethane' – and what makes it a better and cleaner fuel to burn than its chemically-identical twin 'natural gas'? Biomethane is naturally released during the decomposition of organic materials. At the present time, most of the 'biogas' which is released into the atmosphere remains uncontained, and joins all the other greenhouse gases which are released by human activity on the planet, adding to global warming.

When that organic waste is fed into controlled anaerobic digesters – as it now is in many sewage works and a small number of landfill sites – the biomethane captured could not only provide a limitless source of fuel, but actually helps slow down the rising levels of greenhouse gases being released into the atmosphere.

This is effectively 'free' energy going to waste and, in addition to sewage works, other potential sources include the huge amount of food waste and farmyard livestock slurry. Added to that are growing numbers of landfill sites around the world where decomposition of organic waste still occurs unchecked, and the scale of the 'lost' energy is immense. To distinguish it from LNG – Liquefied Natural Gas – it is often referred to as RNG – Renewable Natural Gas – or Natural Renewable Gas, NRG (en-er-gee!). Not only is it a cleaner fuel, it is environmentally beneficial.

When biomethane is ignited, like all hydrocarbons it produces carbon dioxide – but less than half the amount emitted by a similarly-sized diesel engine. This does not add to global warming as it is part of an endless cycle, having been taken out of the air when the organic material from which it is derived was growing. So the net increase of carbon dioxide in the atmosphere is zero – thus it is 'carbon neutral'.

Had this methane not been collected and used as a fuel it would naturally have been released as a global warming gas which scientists have now estimated to be more than eighty times more damaging than carbon dioxide.

LNG is a fossil fuel and is a finite resource. Biomethane, on the other hand, is harvested from infinitely replenishing sources. When conventional fossil fuels are burned, carbon dioxide is released into the atmosphere which would otherwise have remained locked away deep underground had it not been extracted from wells or mines. Biomethane combustion on the other hand, coming from that natural decomposition of organic materials, is part of a natural cycle. Thus, while burning methane extracted from wells adds to the problem of global warming, burning biomethane from recycled organic waste actually lessens it.

Therein lies one of the contradictions in the UK government's proposed future blanket ban on internal combustion engines – it is the type of fuel being burned, not the engine itself, which is the issue.

A computer-generated render of what the proposed railcar could look like.

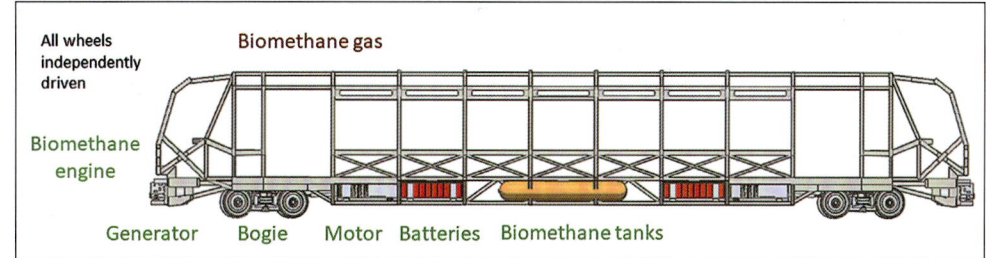

The proposed layout of ULRP's biomethane-engined railcar with the gas tanks and batteries beneath the vehicle's floor.

Another approach – also involving biomethane – proposes resurrecting and updating tramcar technology from more than a century ago, using compressed air. In this proposal, small biomethane-fuelled compressors would top up a tramcar's air tanks at every major stop along its route, so the tramcar itself would have no emissions whatsoever, except for fresh air.

As discussed earlier in this book, a century ago Mékarski compressed air tramcars ran very successfully on some of the tramways in Paris, but the system currently under development is much more sophisticated.

The compressed air engine on board would generate electricity to charge the vehicle's battery packs, and a kinetic energy recovery system using flywheel technology would add to the charging capability during braking. The developers of the system describe it as a 'tribrid' as it brings together the combined benefits of three separate technologies.

But while both of these approaches, biomethane and the compressed air system, have been demonstrated to be both technically and environmentally viable, the major obstacles to be overcome are the vested interests of those committed to battery-electric traction, who have persuaded governments to implement bans on the internal combustion engine whether or not the fuel it uses is a net pollutant. It really has gone full circle, as it was electric traction which eventually brought an end to developing the potential of gas traction a century ago.

Current government policy seems to be being strongly steered towards battery electric vehicles, the increased use of which is dependent on the exploitation of strictly limited natural resources and a ready supply of cheap electricity.

Indeed, in the case of lithium for battery production, its extraction is neither easy nor cheap, and largely depends on very low cost labour in developing countries. Its purification is energy-hungry, requiring the availability of yet more electricity. As to the vehicles themselves, the efficiency of battery-electric trams – like electric cars – is currently limited by the sheer size and weight of the batteries themselves. A considerable percentage of their power usage is expended on the simple task of transporting the heavy batteries along the tramway.

Arguments against pure battery-electric trams or buses draw attention to the very short distances they can travel between charges – making such vehicles heavily dependent on rapid-charging points along their routes.

Another energy source currently under evaluation is hydrogen, the future potential of which was first demonstrated in the 1960s, but it was sidelined by its high development cost and by the dominance of the big oil companies. The dream back then was of an internal combustion engine which burned pure hydrogen, the only emission from which was understood to be water.

Interest in hydrogen technology was only revisited when awareness of the climate crisis became widespread. And while hydrogen is on the verge of a successful future, it is not as a combustible fuel as was envisaged sixty and more years ago, but as a means of in-vehicle generation of electricity through the use of what have become known as hydrogen fuel cells – hydrogen used in this way is not strictly speaking a fuel although widely referred to as such.

Even in its infancy, the technology promises much higher efficiency than petrol or diesel, and with no polluting emissions except those which were created during its manufacture.

In conventional internal combustion engines, the efficiency is low – no more than about 25 per cent of the energy burned is converted to motive power, most of the rest being lost as heat. While less efficient than batteries, even in its development stage hydrogen fuel cell technology is offering more than twice that percentage.

However, while vehicles powered by hydrogen fuel cells are emission free, the production of hydrogen itself is energy hungry and a number of the methods currently in use result in the release of significant amounts of greenhouse gases.

Much of the hydrogen currently available is produced from the 'reforming' of natural gas, and that process uses considerable quantities of electricity and releases significant amounts of carbon. As long as the hydrogen is produced from extracted fossil fuels, it will never be truly 'green'. Producing it from biomass would be a lot greener – another possible use for recovered biomethane – but would probably not realise the quantity of gas which a fully developed hydrogen economy would demand.

The electrolytic extraction of hydrogen from water is also energy hungry. The ultimate goal is to use 'green' electricity generated by either solar, wind, or wave power to isolate the hydrogen by electrolysis, and when that goal is realised, hydrogen power will truly be a sustainable green energy source

Despite the cost and technical challenges, hydrogen fuel cell tramcars are already in production. In China, South Korea and elsewhere, long-term testing of such vehicles in public service is underway, while in Qatar, Dubai and Aruba, vehicles powered by hydrogen fuel cells have already been in public service for several years.

The technology which has made this possible is still evolving. As compression technology allows more hydrogen to be carried on board in an ever smaller space, the range of such vehicles – and the number of passengers they will be able to carry will increase.

The first hydrogen fuel cell tramcars went into service in 2015 on the streets of Oranjestad in the Caribbean island of Aruba. Built by the Californian company TIG/m Modern Street Railway – who also build the hydrogen generators and trackside refueling stations – these pioneering vehicles really do signal an attractive clean and green future for tramways. Each vehicle has its own on-board hydrogen fuel cell array which generates electricity to top up the on-board batteries which, in turn, power the car's electric motors.

Even in its infancy, the technology behind this hybrid approach gives the vehicles an extended operating range.

Aruba, which experiences virtually constant wind and a great deal of sunlight, produces 'green' electricity from its wind turbines and solar farms, and that electricity is used to power the production of hydrogen by electrolysis.

Top: The battery packs used in the vehicles. The system operates at a nominal 400 volts, using LiFePO4 batteries, and TIG/m's own proprietary Battery Management System.

Above: TIG/m's APU (Auxiliary Power Unit) embodies a 16kW self-contained Hydrogen Fuel Cell Generator.

Above left: TIG/m's hydrogen fuelling station. The station comprises a 24kg hydrogen delivery system which handles electrolysis, pumping and compression of the hydrogen to either 350 or 700 bar.

Above right: One of the Dubai tramcars with the iconic Burj Khalifa behind.

Right: Two of Aruba's tramcars – the double-decked hydrogen fuel cell car in the foreground, with a single-deck battery car behind.

The island has now virtually eliminated fossil fuels from its electricity generation portfolio, and the success of the renewable energy fuel cell trams is a 'first' for transportation. The same system was also introduced in Dubai at around the same, using identical vehicles.

Quite apart from the fuel the vehicles use, one of the major drivers in the quest for the ideal future transport system is, of course, cleaner air; but simply replacing petrol and diesel road vehicles with green gas or electric alternatives is only solving one aspect of the problem.

Another of TIG/m's innovative vehicles has been in public service since 2019 in Doha, Qatar, and can operate for up to thirty hours – running over 200 miles – between charging/refueling operations.

Like the company's other vehicles, it uses battery-dominant hybrid propulsion architecture, but in a nod to future adaptability, it can be set up to use a range of on-board auxiliary power units and fuels. In one configuration it might have an internal combustion engine using biomethane, CNG, or LPG, while in another a hydrogen fuel cell might be used. These could be interchanged in a matter of hours.

At the time of writing, with the growing interest in the creation of city-centre 'clean air zones', a number of British towns and cities are exploring the suitability of such systems for reducing pollution without the huge cost of overhead electric catenary.

Irrespective of the chosen technology, one factor which is growing in significance with urban transport decision-makers concerns the potential replacement of buses and cars in town centres with tramways.

There is growing concern about the microscopic rubber/plastic beads created by the wearing down of vehicle tyres as they are abraded by the road surface. Running steel wheels on tramway rails, however, eliminates that issue – another reason why the tramcar should figure strongly in future transport strategies.

Above: Hybrid fuel cell Car No.1 on the Dubai street tramway.

Left: One of the award-winning hybrid trams operating in Doha, Qatar. This vehicle uses a hydrogen fuel cell, but could also be fitted with a biomethane-burning internal combustion engine. Versions are currently under consideration in other towns and cities, and a light rail option was demonstrated in the US in October 2021. For city centre tramways, the hydrogen fuel cell option is the cleanest, while for urban tram and rail use, biomethane offers a substantial reduction in pollutants over diesel.

SELECTED PATENTS 1791–1903

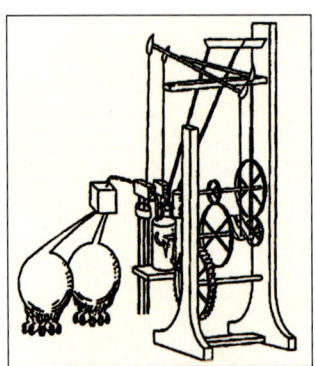

The illustration of John Barber's gas and steam turbine engine which accompanied British Patent No.1,833 from 1791.

What is believed to have been the first patent for a gas engine was granted to John Barber (1734–1793) in 1791 for a gas turbine which he envisaged being used to power road vehicles amongst other things. Between then and 1860 when Lenoir was granted his first patent, several dozen other designs for gas engines had been patented across Europe.

Of more than 700 related patents filed across the world in the half century after 1860, many were little more than slight modifications and improvements on earlier patents by the same engineers. Figures such as Nicolaus Otto, Carl Lührig, the Crossley Brothers and Percy Holt must have spent very large sums of money filing a new patent for even very minor changes.

Others who, as far as this research has been able to determine, never actually built and tested the tramcars for which they sought patent protection, nonetheless filed multiple patents across the world's major engineering nations – Great Britain, Belgium, Germany, Italy, France, Switzerland, Australia, New Zealand, the United States and Canada.

Between 1879 when the first patent was registered for a gas-engined tramcar, and 1920 when the Neath trams were withdrawn from service, there were more patents granted for gas-engined tramcars than there were gas tramcars built.

The lack of scrutiny by the patent authorities round the world which was commonplace at the time – and the absence of today's sophisticated communications – meant that vague claims to have invented many of the key elements of a successful gas-engined vehicle have been protected by multiple patents in the names of several people.

1791, Acc. *31 October*: GB Patent No.1,833
John Barber:
Obtaining and Applying Motive Power, &c. A Method of Rising Inflammable Air for the Purposes of Procuring Motion, and Facilitating Metallurgical Operations. Believed to be the earliest patent for a gas engine, air and gas were to be compressed in individual cylinders and pumped into a combustion chamber which he called an 'exploder' where they were ignited, the hot exhaust gases thus produced being directed against the vanes of a paddle wheel. Water was to be injected onto the explosive mixture as it left the 'combuster' producing steam, to increase the effect. While based on sound principles, it produced insufficient power to compress the gas and air and thus was incapable of useful work.

1838, App. dated *18 April*: Acc. *15 October*: GB Patent No.7,615
William Barnett:
Certain Improvements in the Production of Motive Power. Barnett's Patent is interesting on a number of counts – firstly its early date and secondly that he contemplated primarily using hydrogen, or 'carburetted hydrogen' – an obsolete name for methane – rather

than town gas or producer gas as the primary fuel. Thirdly, he was allowed to patent the use of those gases 'to give motion to machinery by their explosive force within a cylinder, or other close vessel, and whether producing a rectilinear or rotatory motion. Secondly I claim the employment of pumps to supply the requisite quantities of air and inflammable gases, and to regulate their several proportions, whether such air and gas be delivered by the pumps directly into the cylinder, or into an intermediate vessel; but I do not claim nor limit myself to the particular construction of pumps shewn in the Drawings.' He also included a general cover-all clause 'And I further declare that the mechanical arrangements, herein-before described, are illustrative and explanatory only, and that I do not intend in all cases to limit myself thereto.'

1853, App. *30 November*: GB (Provisional) Patent No.2,789
Alphonse Loubat:
Improvements in the Construction of Tramways. Loubat is credited with developing the profile of the tramway girder rail which is still in widespread use across the world today. This application was sub-headed 'Application of a Grooved Tram or Rail to Common Roads', using the original definition of the word 'tram'. However, in Britain he only filed a Provisional Specification. Summarised as 'This invention relates to the application to ordinary roads of a rail having a grooved upper surface, such rail being intended to serve as a tramway for waggons and other vehicles. The rails are set in such a manner in the road as not to project above the surface thereof; they will therefore present no obstacle to the progress of the ordinary traffic.'

1856, Acc. *7 May*: GB Patent No.1,071
William Joseph Curtis:
Improvements in Carriages to Run on Rail or Tram Ways and Common Roads. Summarised as '… in order to cause the carriage wheels when running on tram or rail ways to keep on the rails, there are additional smaller wheels applied to the carriage, suitable for running in or on the rails, and these additional wheels are capable, by levers and connecting rods or suitable apparatus, of being raised and lowered by the driver or other person, hence, when it is desired that the carriage shall be retained on the rails, the additional wheels shall be lowered and become guide wheels for the carriage; but when the carriage is run off the rails on to a common road, the additional wheels are raised and kept up out of the way, and such is the case so long as the carriage is to run on a common road.'

William Joseph Curtis's proposal for a hybrid vehicle which could operate as a two-horse coach on open roads, but could also operate on a tramway by the lowering of flanged wheels. His patent does not address the challenge of aligning the vehicle with the tramlines before lowering the wheels. The raising and lowering of the wheels was to be performed manually by the driver or 'his attendant' using what is simply described as 'suitable apparatus'. When the flange wheels were lowered, they were to be locked together with the road wheels. Those complex tasks – aligning the car with the tramrails and lowering and locking the wheels – could only have been undertaken when the car was empty as there were no hydraulics to make the job easier. There do not appear to be any recorded instances of the vehicle being tried on any British tramways.

1860, App. *24 January* French Patent No.43,624, & *8 February* GB Patent No.335: GB Patent Acc. *8 August 1860*
Jean Joseph Etienne Lenoir:
Improvements in Obtaining Motive Power and in the Machinery or Apparatus Employed Therein. The first of several Lenoir patents, actually patented in France a week earlier, was summarised by his Patent Agent John Henry Johnson – 'This invention consists in the application and use of an inflammable gas mixed with a proper proportion of atmospheric air and lighted inside a cylinder by the aid of electricity, the expansion thereby produced acting upon the piston and imparting motion thereto which motion may be transmitted in any convenient and well known manner to a driving shaft.'

1862, App. *16 January*, Acc. *10 March*: French Patent No.52,593
Alphonse-Eugene Beau de Rochas:
Nouvelle Recherches sur les Condition Pratique de Plus Grande Utilisation de la Chaleur et en Général de las Force Motrice. A supplemental claim was applied for on 10 June and accepted on 26 August. This document lies at the heart of a major nineteenth century dispute over who first proposed the four-stroke engine cycle, later to become known as the 'Otto-cycle'. A lithographed of his hand-written text survives in which he describes 'the best employment of the elastic forces of gas' and how to achieve them, listing the stages as suction during the entire first stroke of the piston, compression during the following stroke, ignition followed by expansion on the third stroke, and exhaust on the final stroke. In France, Beau de Rochas failed to pay the required 100 francs annual fee, thus causing his application to lapse after just one year. In 1885, the Körting Brothers successfully used the Beau de Rochas patent application to get part of Otto's claim to have invented the four-stroke engine removed from his German Patents. In Britain, however, Otto's primacy for the idea was upheld.

1872, App. *7 June*, Acc. *7 June 1872*: GB Patent No.1725
Henry Percy Holt:
Improvements in Machinery and Locomotive Engines for Driving Tram Cars and other Road Vehicles, Parts of which Improvements are Applicable to other Steam Engines. Before developing his interest in gas-engines trams, Holt designed and patented a small articulated steam locomotive which could replace the horse on an 'ordinary' tram car.

1872, App. *20 December*, Acc. *21 February 1873*: GB Patent No.3869
Henry Percy Holt:
Improvements in Machinery and Locomotive Engines for Driving Tram Cars and other Road Vehicles, Parts of which Improvements are Applicable to other Steam Engines. Improvements to his articulated steam locomotive design.

1876, App. *20 December 1875*, Acc. *23 May 1876*: US Patent No.177,736
Louis Mékarski:
Improvements in Devices for using Compressed Air for Motive-Power. The American version of Mékarski's original French patent for his compressed air tramcar engine. In it he offered two different approaches to stopping the air chilling and icing up the cylinder as it decompressed. In the first, a small coal-burning heater was

employed, while in the second version, a jacket around the cylinder was filled with near-boiling water each time the car was refilled with compressed air. He claimed that 'The most important result of this invention is the possibility of storing the air in carriages at a very high pressure, (twenty-five atmospheres or higher, permitting a long journey without recharging the reservoirs.' He acknowledged that long journeys required multiple compressed air cylinders, which considerably increased the tramcar's weight.

1876, App. *10 January*, Acc. *30 May*: US Patent No.178,023
Nicolaus August Otto:
Improvements in Gas-Motor Engines. Summarised as 'In gas-motor engines wherein a combustible gaseous mixture is exploded without doing work, and the contraction of the products of combustion due to cooling is utilised for producing motive power— 1. The method of alternately arresting and permitting the movement of a piston by regulating the flow of a liquid in a hydraulic cylinder, substantially in the manner and for the purposes herein set forth: 2. The combination of the loose piston with a piston working in a hydraulic cylinder provided with passage and valve, operated by a cam, substantially as herein described.' Largely similar to German Patent No.532 of 5 June 1876 and 4 August 1877.

1876, App. *17 May*, Acc. *1 August 1876*: GB Patent No.2,081
Nicolaus August Otto:
Improvements in Gas-Motor Engines. Believed to be the earliest patent granted in Britain which fully articulated the four-stroke engine cycle. Summarised as 'First. Admitting to the cylinder a mixture of combustible gas or vapour with air separate from a charge of air or incombustible gas to that the development of heat and the expansion or increase of pressure produced by the combustion are rendered gradual… Second. Compressing by one in stroke of the piston a charge of combustible and incombustible fluid drawn into the cylinder by its previous out stroke, so that the compressed charge when ignited propels the piston during the next out stroke, and the products of combustion are expelled by the next in stroke of the piston…'

1876, App. *13 July*, Acc. *14 August 1877*: US Patent No.194,047
Nicolaus August Otto:
Improvements in Gas-Motor Engines. Summarised as '1. A gas-motor engine wherein an initial mixture of combustible gas or vapour and air is introduced into the cylinder, separate from a charge of air or other incombustible gas, in such a manner and in such proportions that the particles of combustible mixture will be close together at the point of ignition, but will be more and more dispersed in the charge of air forward of that point, whereby the development of heat and the expansion or increase of pressure produced in the combustion are rendered gradual, substantially as herein described.' The patent also claimed protection for many aspects of the engine's design.

1877, App. *26 July*, Acc. *23 October*: US Patent No.196,473
Nicolaus August Otto, Francis W. Crossley and William J. Crossley
Improvements in Gas-Motor Engines. The US version of their British Patent filed on 4 June 1877. Also patented in Canada on 16 October with a handwritten Specification

as Canadian Patent No.8023. An evolution of US Patent 194,047 summarised as 'Now, according to our present invention, instead of producing a stratified condition of the combustible charge, such as is obtained by the means last above described, we effect a similar gradual development of the motive power by introducing into the cylinder a uniformly-diluted mixture of gas and air—that is to say, a mixture containing a greater proportion of air than the strong or explosive mixture which only contains as much air as is necessary for the complete combustion of the combustible gas—so that on ignition this weak charge will burn with considerably less rapidity than the explosive mixture.'

1878, App. 11 January 1878, Provisional GB Patent No.10
Mathew Hilton and James and Samuel Johnson:
Improvements in in the Application of Gas Motors to Tram Cars and Other Self-propelling Vehicles. This provisional specification contained many of the aspects which Lührig and others would later also patent for their tramcars. Two unique propositions were that the gas could be stored in weighted telescopic holders, and that the tramcar engine might be used to compress the gas as it was taken on board.

1878, App. 13 December 1878, Acc. 31 May 1879: GB Patent No.5113
Francis William Crossley and William John Crossley:
Improvements in Gas Motor Engines. Summarised as 'First. In a gas motor engine, the cylinder of which is open to the atmosphere at one end, causing the hot gases resulting from the combustion of the charge during one outstroke of the piston to pass through a passage provided with a check valve into a receiver, from which the stored gases are subsequently admitted into the cylinder to cause the piston to perform its next outstroke, substantially as herein described. Second. In combination with a gas motor engine operating as set forth in the preceding claim the use of a revolving slide with ports and passages so arranged as by one rotation consecutively to introduce the combustible charge into the cylinder to fire the charge, to admit gas from the receiver to the cylinder, and to exhaust the gases from the cylinder to the atmosphere. Third. In a gas motor engine, the cylinder of which is open to the atmosphere at one end, causing a combustible charge introduced into the cylinder during the first part of the stroke to be expanded during the remainder of the stroke and then fired, the gaseous products of combustion being made to escape through a passage provided with a check valve, so that a partial vacuum is produced under the piston thereby causing this to perform its instroke.'

1879, App. 13 May (Full Specification never submitted), GB Patent No.1912
Henry Percy Holt and Francis William Crossley:
Improvements in Machinery for Starting, Propelling, and Stopping Vehicles, and in the Apparatus and Appliances Connected therewith, more Particularly with Reference to Gas Engines and Tramways, but also Applicable with other Motor Engines. A general summary of the operational features of their proposed tramcar, but without specific detail.

1879, App. 25 January , Acc. 22 July 1879: GB Patent No.309
Conrad Krauss:
A Gas-power Locomotive for Tramways and for Railways of Secondary Order. Summarised as 'I do not claim all the details described in the foregoing Specification, as they

may be varied without departing from the principle of the Invention, but I claim as new and essential features of the same:– 1. In a locomotive or carriage driven by gas-power, the storage of compressed gas in one or more reservoirs, which gas, after being reduced to a certain lower but constant pressure by means of a pressure-regulator, is mixed in adjustable proportions with air of the same pressure, supplied by an air-compressing pump driven by the engine; the gas and air thus mixed being thereupon conducted to the gas-motor, substantially as described. 2. The transmission of motion from the gas-motor to the axles of the locomotive or carriage by means of friction-wheels, friction-couplings, or other frictional gearing for the purpose of allowing the locomotive or carriage to be started or stopped, or its motion reversed while the gas-motor continues working with its shaft, always revolving in the same sense. 3. The design of the gas-motor in combination with a locomotive or carriage ... in which two pistons acting in opposite directions in one cylinder operate on two coupled crank-shafts provided with friction-wheels either of equal or different diameters...'

1879, Acc. 29 October 1879, Acc: 23 April 1880: GB Patent No.4306
John Roger Purssell:
An Improved Arrangement of Apparatus for Moving Tram Cars and other like Vehicles by Gas Engine Power. A gas tramway locomotive unit fitted with two single-cylinder gas engines mounted side-by-side and acting upon a common shaft, summarised as 'An apparatus for moving tram-cars, the combination of a gas-engine, a transverse horizontal shaft connected at one end with the piston of the engine, a worm arranged on the shaft, a horizontal worm-wheel engaging the worm and rotating it continuously in one direction, and a vertical shaft carrying the worm-wheel at its upper end and provided at its lower end with means of rotating and driving axle of a car in reverse directions, as required.'

This was a revised version of a Provisional Specification (No.1727) he had submitted a few months earlier. The vehicle was also described and patented in the United States as US Patent No.231,097 dated 10 August 1880.

1879, App. 4 November, Acc. 30 April 1880: GB Patent No.4499
Henry Percy Holt and Francis William Crossley:
Improvements in Machinery, Apparatus, for Starting, Propelling, Stopping, and Reversing the Direction of Motion of Vehicles on Rails, Tramways, or Roads, more Particularly Applicable to Gas Engines, but also Suitable for other Motor Engines. Summarised as 'The use for starting, propelling, stopping and reversing the direction of motion of locomotive vehicles, of the machinery, apparatus, and appliances, substantially as herein described, whereby the driving axle can be lowered or raised, and turned to an inverted position relatively to the vehicle, so as to vary the grip of the driving wheels, and to stop or reverse the motion of the vehicle without stopping or reversing that of the engine which drives it.'

1879, App. 27 October, Acc. 9 December: US Patent No.222,467
Gottlieb Willhelm Daimler:
Improvement in Gas Motor Engines. Summarised as 'A gas-motor engine having two parallel and disconnected high-pressure cylinders, each of which consecutively draws in a combustible charge, compresses it, fires it, and performs the working-stroke, and expels the products of combustion, the two disconnected and independent

pistons of the two cylinders being connected to one and the same engine shaft, so as to perform their strokes together, but the working-stroke of the one cylinder being made to take place while the other cylinder is drawing in its combustible charge, substantially as described. 2. In a gas-motor engine, the combination of two high-pressure cylinders, each of which consecutively draws in a combustible charge, compresses it, fires it, and performs the working-stroke, with a low pressure cylinder into which the gaseous products are caused to pass alternately from the high pressure cylinders, so as to perform further work by expansion therein, substantially as described.'

1881, App. 20 December, Acc. 29 June 1882: GB Patent No.5575
Joseph Quick and Joseph Quick the younger:
Improvements in Tramway Locomotives and other Locomotives or Motive Power Engines. Summarised as 'First:- The combination substantially as herein described in a locomotive engine, of a gas reservoir, gas engine, and frictional clutching arrangements, through which to drive the wheels, such combination admitting of the gas engine being maintained at a constant or nearly constant speed whether the locomotive be running or at rest or starting or stopping. Second:- The combination substantially as herein described in a locomotive engine of a high pressure gas reservoir, a gas engine, and pumps to be worked by the said engine to charge the reservoir. Third:- The combination substantially as herein described in a locomotive or other motive power engine, of a high pressure gas reservoir, a low pressure service reservoir, a gas engine, and a water circulation withdrawing heat from the cylinders and imparting it to the gas during expansion on passing out of the high pressure reservoir.'

1882, App. 15 February, Acc. 13 June 1882: US Patent No.259,413
Jean Marie Armand Montclar:
Gas Loco-motor for the Locomotion of Vehicles, Carriages, Tramways, Wagons &c. Introduced as 'The application of gas-motor engines and the special arrangements combined therewith to all kinds of vehicles for the purpose of substituting the same for horses or other beasts of burden, as well as for steam and compressed air engines hitherto used for the locomotion of carriages, omnibuses, fire-pumps, tramway-cars, railways &c.' and summarised as 'The combination of a gas-motor engine, having one or more cylinders, and main crank-shaft, with which the pistons of said cylinders connect, with an air pump operated by the said crank-shaft to assist in starting the motor, and adapted to operate the same…'

1882, App. 28 August: New Zealand Patent No.681
Walter Andrew Harper and John William Rock:
A new Gas Engine which may be used as a Locomotive, especially in Propelling Tram-cars. Summarised as 'The construction of mechanism for applying power to drive street-cars, to means of applying the brakes to power-driven street-cars, and also the means for counteracting the disagreeable noise and odours from gas-engines and similar motors when applied to street-cars as a motor.'

1883, App. 29 August, Acc. 27 February 1884: GB Patent No.4165
Thomas Fothergill McNay and Frederick James Harrison:
Improvements in the means of utilizing Gas Engines for locomotion. The design sought to avoid the 'jolt' when the clutch was activated on gas engines, by using gears and

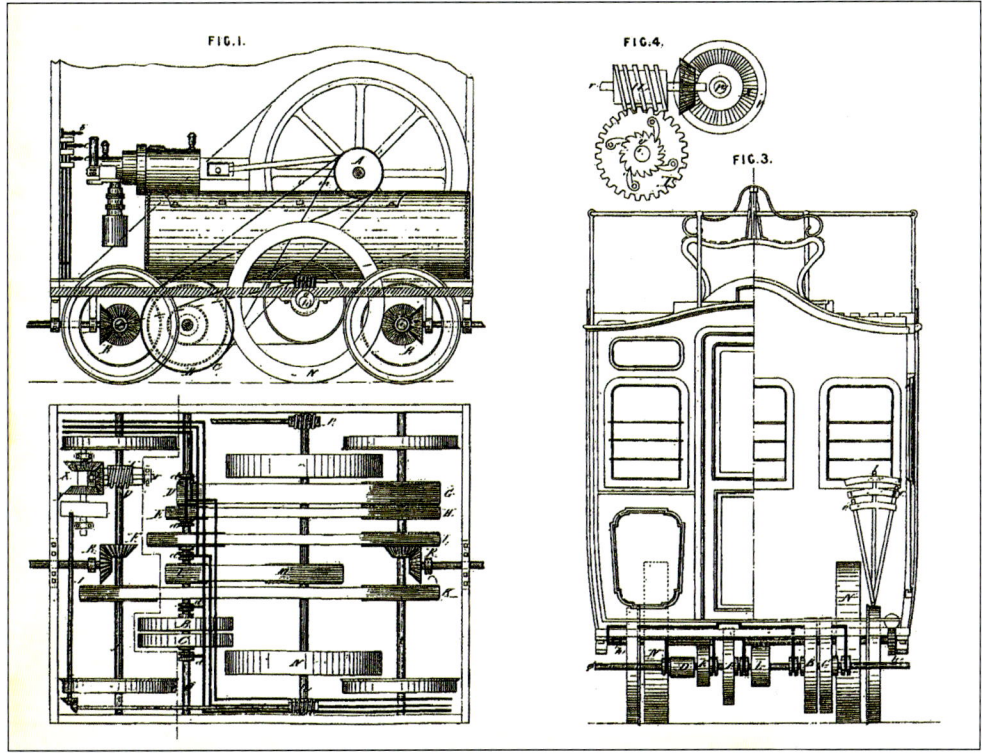

Thomas McNay and Frederick Harrison seem to have been proposing a unique hybrid drive system which, in addition to the wheels which ran on the tram rails, two additional powered wheels would run on the road surface itself, the driver employing a pulley system of gears to control the power going to those wheels. They envisaged the gas supply being contained in a large tank on the vehicle, augmented by a compressed air supply which would keep the engine turning when stationary. There seems to be no surviving record of their unusual tramcar design being built and tested.

pulleys to smooth the vehicle when moving off from stationary. The specification also suggested 'all wheel drive' for improved traction and control.

1884, App. 25 April, Acc. 23 January 1885: GB Patent No.6784
Andrew McNeill:
Improvements in Tramway Locomotives Driven by Gas. An ingenious Otto-engined gas powered locomotive containing its own in-built gas generation using any one of a range of mineral oils as fuel, the gas generator being itself driven by the Otto engine. There is no evidence the vehicle was ever built.

1884, App. 25 November: Victoria (Australia) Patent No.3,882
John Danks and Benjamin Barnes:
Improvements in Tram Car Motors. Summarised as 'First: The combination and application of a gas engine with gas (or gas and air) stored under pressure and carried in suitable vessels or receivers on (or attached to) the car for working the gas engine as a tram car motor. Second: The Combination and arrangement of a tram car, gas engine and receivers as hereinbefore described with the arrangement or friction gear for transmitting the necessary rotary motion to the wheels of the vehicle, and for effecting (by the aid of levers) the operations of starting, stopping and reversing the motion of the tram car without stopping the engine.' As Australia was, at the time, made up of autonomous colonies, a separate patent was registered in South Australia.

1884, App. 27 December, Acc. 3 September 1885: GB Patent No.16,947
John Danks and Benjamin Barnes:
Improvements in Tram Car Motors. Essentially the same specification as their Australian patents.

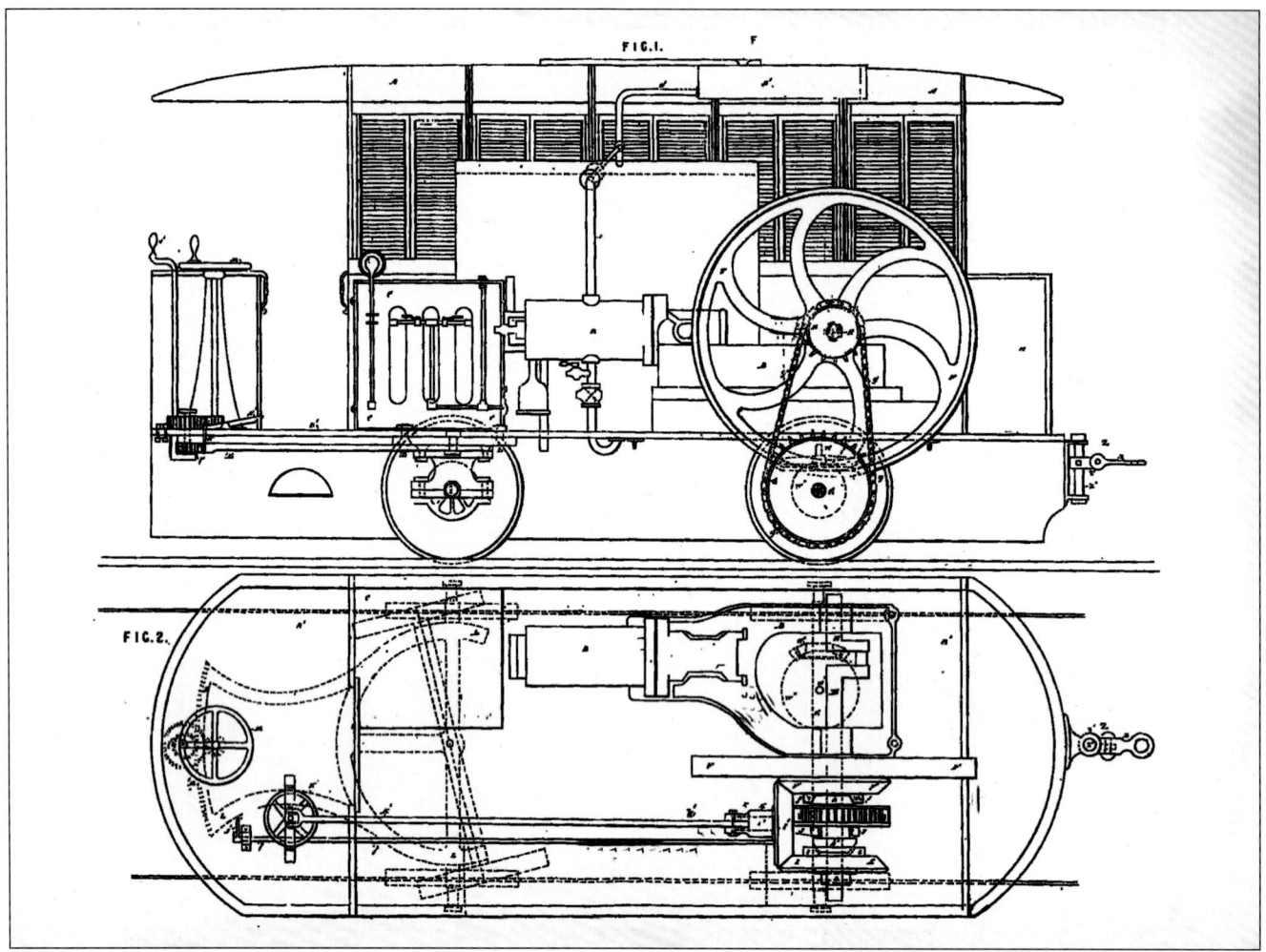

This gas-engined tramway locomotive – with its built-in gas generator – was designed and patented in 1884 by Andrew McNeill from Liverpool, who described himself as a 'Shipbuilder and Engineer'. There is no record of it ever having been built. One innovative feature was the suggestion that a pivoting front axle might be incorporated to facilitate the negotiation of sharp curves.

1885, App. 7 February, Acc. 23 June: US Patent No.320,634
John Danks and Benjamin Barnes:
Tram-Car Motor. Summarised as 'The combination, substantially as described, with a street car provided with seats for passengers, storage-tanks for storing a motive power under pressure, a gas-engine connected with said storage-tanks and mounted on the car, and one of the wheel axles of said car, of transmitting mechanism for transmitting motion to the wheel-axle, consisting of the friction wheel on the engine-shaft, the friction wheel on a counter or intermediate shaft, the chain-sheave on said shaft, the chain-sheave on one of the axles of the car, and the driving-chain substantially as and for the purpose specified.' Although the language differed slightly, this was the US version of Victoria (Australia) Patent No.3,882.

1886, App. 19 March 1886, Acc. 13 July 1886: US Patent No.345,279
John S. Connelly and Thomas E. Connelly:
Street-Car Motor. A design for a chain-drive gas-engined 'street-railway car' with an integral motor and fly-wheel, intended to enable a smaller engine to be used to drive the tramcar. The key claim within the patent was that 'The continuous revolution of the fly-wheel while the car is stopped stores sufficient energy to overcome the inertia of the car to start it with ease.' This claim would subsequently also be key to Lührig's patents, but he would employ a much larger fly-wheel.

1886, App. 7 May 1886, Acc. 17 August 1886: US Patent No.347,470
John S. Connelly:
Car Motor. An evolution of the car specified in the previous patent, but with the important addition of a rudimentary gearbox giving the engineman four gears to select from as the vehicle moved off from stationary. Again a small fly-wheel is employed.

1886, App. 15 November 1886, Acc. 18 September 1887: GB Patent No.14,802
Emmanuel Stevens:
Improvements in Apparatus or Machinery for Driving Tramway and other Carriages by Gas, Hydrocarbon, or other Portable Motors. Improved clutch and gearing, summarised as 'such gearing being especially suitable for use with gas, spirit or other portable motors, whereby the said tramway or other carriage or car may be driven either forwards or backwards or be caused to remain entirely stationary, and the speed varied, without stopping or re-starting the motor.'

1886, App. 25 November 1886, Acc. 25 October 1887: GB Patent No.15,359
Richard McGhee and John Magee:
Improvements in Motor Cars or Vehicles Propelled by Petroleum Vapour or by Gas for use on Railways, Tramways, and for General Traction Purposes. A gas-engined tramcar which

The Belgian engineer Emmanuel Stevens (spelled Emanuel in some of his patents) designed his gas tramcar with four horizontally mounted gas storage tanks taking up one side of the vehicle. He suggested that the same general design could be adapted to use 'spirits or other fuels' as well as gas. There are no records of the vehicle being built.

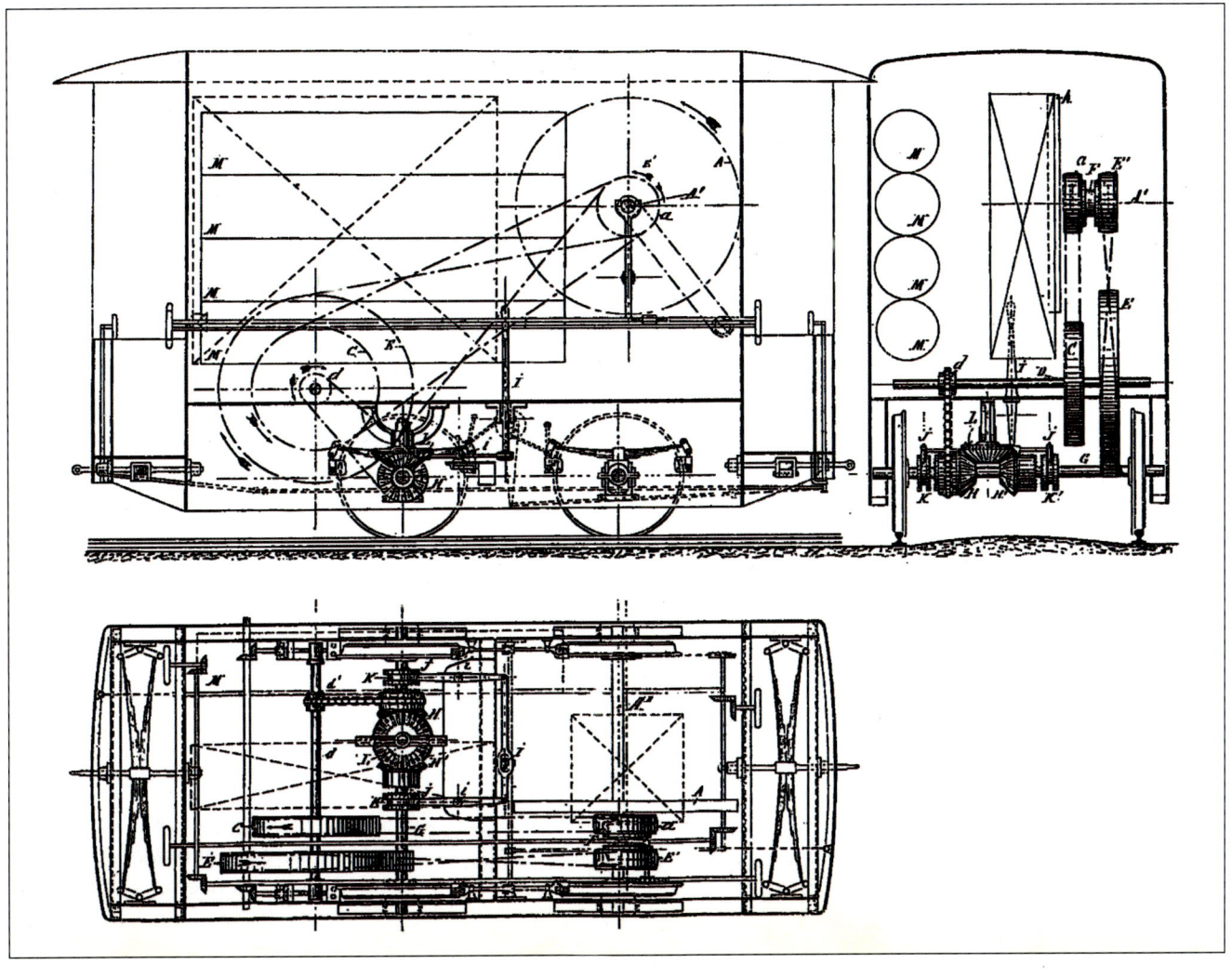

190 • THE GAS TRAMCAR

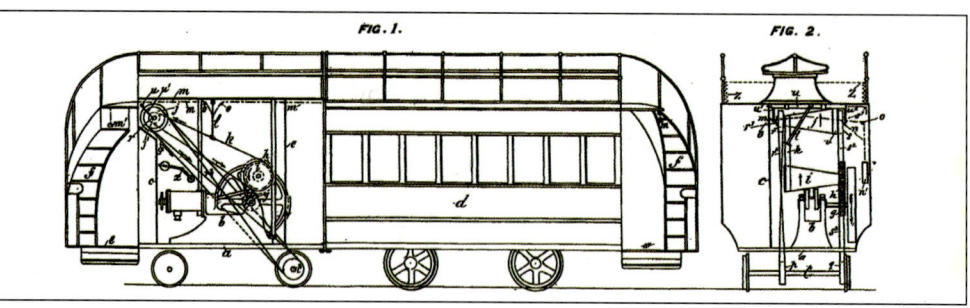

Right: Richard McGhee described himself as a 'Commercial Traveller' on his patent specification while John Magee – different spelling but resident at the same address in Shawlands, Glasgow – described himself as an 'Engineer'. The length of this vehicle would have proved impossible to run on the tight curves of Glasgow's tramways. It is assumed that it was never built.

Below: Emanuel Stevens' 1887 Patent, for a radically different vehicle to that patented in the previous year, seems to have been yet another which never got far beyond the drawing board, despite also being patented in the Unites States. The illustration with US Patent No.414,173 – *bottom*, was usefully annotated identifying the function of each element of the design.

could be driven from either the driving position next to the engine, or from linked controls at the opposite end of the passenger car. The inventors also specified a 'speed cone' gearing system to enable the driver to control the vehicle's speed, and compressed air brakes. The diagram suggests they planned to use a small 'off the shelf' portable gas engine to power their vehicle.

1887, App. *31 March 1887*, Acc. *29 March 1888*: GB Patent No.4843
Emanuel Stevens:
Improvements in Apparatus or Machinery for Driving Tramway and other Carriages by Gas, Hydrocarbon, or other Portable Motors. While having the same title as Stevens' 1886 Patent, this referred to a hybrid gas/compressed air engine. Summarised as 'The combination of an air motor having a double walled cylinder, a gas motor likewise provided with a double walled cylinder and adapted to produce compressed air for accentuating the air motor and a refrigerator in which cold is produced by expansion of exhaust gases from the air motor.' An innovative feature was the inclusion of 'silent exhaust apparatus' to minimise noise while in operation.

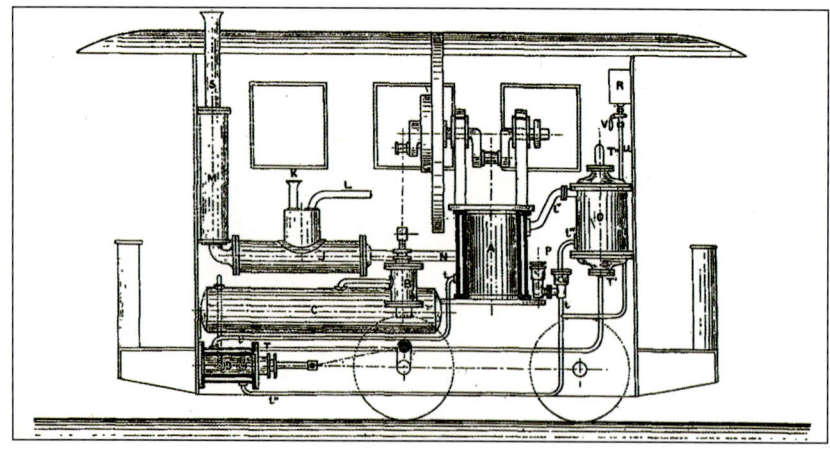

1887, App. *26 April 1886*, Acc. *6 September 1887*: US Patent No.369,610
Jay Noble:
Power-Driven Street Car. Summarised as 'The construction of mechanism for applying power to drive street-cars, to means of applying the brakes to power-driven street-cars, and also the means for counteracting the disagreeable noise and odours from gas-engines and similar motors when applied to street-cars as a motor.'

1887, App. *21 September 1887*, Acc. *15 May 1888*: US Patent No.383,065
Nicolaus August Otto:
Motor Engine Worked by Combustible Gas. Also patented as GB Patent 11,503 23 August 1887; in Belgium, Patent No.78,765 3 September 1887; in Italy Patent No.176 10 November 1887.

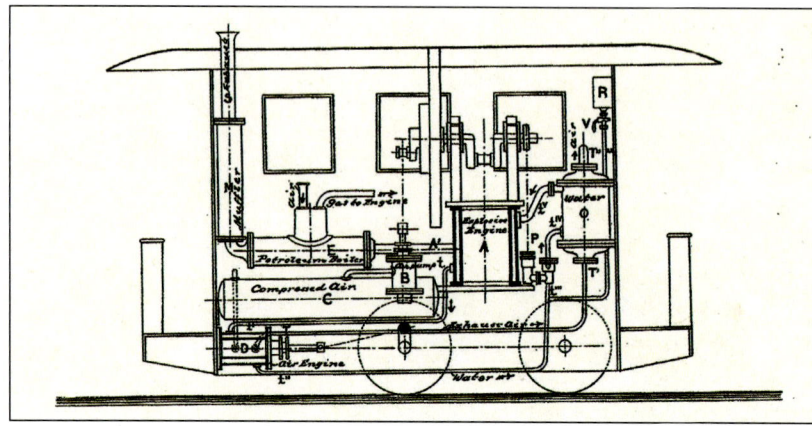

Summarised as 'In gas-motor engines working with a compression space and a cycle of four strokes, the method for replacing the residual products of combustion remaining in the cylinder at the end of the expelling-stroke by a combustible charge by first driving out the said products by means of a charge of atmospheric air and then drawing in a variable quantity of combustible gas with or without admixture of air to mix with the said charge of air, the combined charge being partially expanded below atmospheric pressure during the suction-stroke and compressed during the return-stroke.'

1888, App. *13 February*, Acc. *14 April 1894*: US Patent No.391,774
Oskar Blessing:
Tram-Car. Summarised as in 'tram-cars or other like vehicles driven by gas or petroleum motors; and the improvements have for their purpose, first, to prevent on starting and reversing of the car for forward as well as for backward movement, a sudden reaction on the motor and a stoppage of the latter thereby; second, a form of clutch for speed-gearing and other purposes devised for normal disengagement; third, to allow of a forward and rearward movement of the vehicle without reversing the motor.' The vertical Blessing engine caught the attention of Carl Lührig who suggested its suitability for powering gas trams in his September 1891 Patent No.15,655, although the drawings with that patent illustrate a larger and more sophisticated engine.

1888, Acc. *15 August*: German Patent No.45,129
Emil Capitaine:
Improvements in Gas Motors. Capitaine's design for a water-cooled gas engine was also patented in the US as Patent No.406,160, and it was reported in 1896 to have been tested in a tramcar, however no further information about such a vehicle has been forthcoming. Capitaine is best remembered for his 1897 law suit against Rudolf Diesel over the primacy of the invention of the diesel engine, and for his early motorcycles.

1890, App. *7 May 1891* Acc. *23 July 1892*: US Patent No.426,985
John S. Connelly:
Improvements in Car Motors. After patenting several designs for a tramcar with an integral engine which could run on gas or oil, this was his first to specify a separate locomotive car powered by a gas engine and fuelled by air charge with 'the vapor of a liquid volatile substance—such as naphtha or gasoline'. An evolution of this vehicle, driven by a two-cylinder engine and built by Weyman of Guildford, was given a six-month trial in London and Croydon.

1891, App. *20 May 1891*, Acc. *10 October*: US Patent No.8652
George Benjamin Nichols and William March:
Improvements in Gas Locomotives for Drawing Tram Cars and Railway Carriages as they are at present constructed. After initially submitting a specification on 16 February 1891, (Provisional Patent No.2765 *Gas Locomotives for Drawing Tram Cars*), Nichols and March declined to complete the process, instead issuing this new specification. Their novel patent essentially proposed a powered version of the horse tram patented by William Curtis in 1856, whereby the tramway wheels could be raised to allow the vehicle to leave the rails and run on an ordinary roadway. This was to be facilitated as follows:- 'To the end of the locomotive is fixed a combined lifting jack and steering wheel by which when the line is blocked by anything in front of it the locomotive can be raised and steered off the rails and guided in any direction

192 • THE GAS TRAMCAR

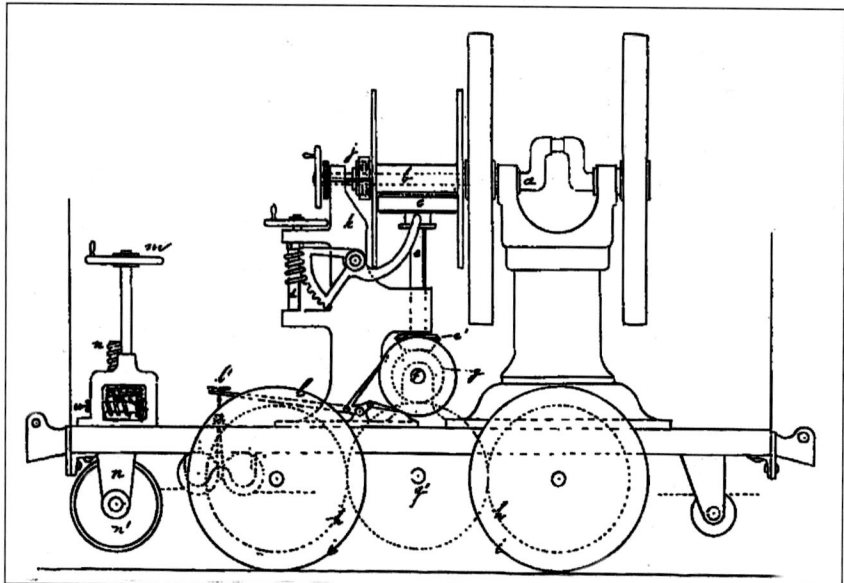

George Benjamin Nichols, a Civil Engineer, and William March a Mechanical Engineer, both of South Hampstead, were granted Patent No.8652 on 10 October 1891 for their vehicle designed to run on either road or rail. They had earlier filed a provisional patent specification for a very different locomotive (Provisional Patent No.2765, 16 February 1891) but did not complete the paperwork for the full specification.

forward or backward. It also proposed using friction clutches and variable gearing to enable the engine speed to be altered according to the demand being placed on it – viz, high speed and low gear for starting, reducing speed for normal running.

1891, App. 15 September, Acc. 23 July 1892: GB Patent No.15,655
Carl Lührig:
An Improved Construction of Locomotive Vehicles. Summarised as 'In locomotive vehicles the use of two or more motor engines so arranged that their motion in one and the same direction is transmitted independently to one and the same shaft or wheel axle, through ratchet gear or equivalent mechanism.' The specification also included an exhaust muffler, or silencer, and provision for using the engine's own waste heat to warm the passenger saloon.

1891, App. 3 December, Acc. 6 August: GB Patent No.21,121
Carl Lührig:
An Improved Construction of Locomotive Vehicles. Summarised as 'In a locomotive vehicle constructed as described in application No.15655 the use of mechanism whereby on turning a hand wheel the brakes are actuated and simultaneously the power of the motor engines is either transmitted to the wheel axles or such transmission is interrupted, while at the same time a braking action can be applied to the engine regulators…'

1892, App. 23 March, Acc. 14 Jan 1893: GB Patent No.11,506
Carl Lührig:
Improvements in Tramway Vehicles Driven by Motor Engines. Summarised as 'In tramway vehicles driven by motor engines, mounting the motor engines under the seats of the vehicle upon a frame, by which they can be removed from the outside on turning up a flap-like portion of the side, while the entire driving gear… is mounted as a whole on two bearers by which it is fixed from below to the carriage frame, whereby the engines and gearing are made separately readily exchangeable, and the penetration of smell and noise to the interior of the vehicle is prevented, substantially as described.'

1892, App. 27 December 1890, Acc. 21 June 1892: US Patent No.477,444
James Morris O'Kelly:
Improvements in Tram-Cars and Motors Therefor. Kelly's patent proposed installing a small vertical single-cylinder gas engine to power a tramcar, but offered nothing specific as to the engine's configuration. Indeed, the inventor supplied more detail on the quality of the seating and the provision of retractable parasols for upper deck passengers than he did on the engine. As far as can be ascertained, it was never built.

1892, App. 25 June, Acc. 1 October: GB Patent No.11,897
Carl Lührig:
Improvements in or relating to the Wheels and Axles of Tram-cars. Summarised as 'A wheel and axle connection for tramcars in which provision is made for the independent acceleration or retardation of motion, for the simultaneous longitudinal displacement of the wheels a sleeve or clutch mounted upon the axle being for the purpose connected by means of spring-controlled stops or pawls with the loose hub or boss of the wheel...' intended to smooth travel around tight curves.

1892, App. 3 September, Acc. 8 October: GB Patent No.15,841
Carl Lührig:
Improvements in Tramway Vehicles Worked by Engine Power. Two of the nine claims summarised as '2. In a carriage such as is above referred to the arrangement of the motor engines under the longitudinal seats, of the fly wheels in the hollow side walls, of the gas-collectors in the front walls, of the driving gear below the floor, and the construction of the seats, the protective casings, the side and front walls and the floor so that they can be folded up or taken to pieces with doors or flaps which can separately be opened, for the purpose of rendering them easily accessible for inspection or repair. 3. In a carriage such as is above referred to the arrangement of the motors upon a moveable under-plate for the purpose of withdrawing them from the carriage and substituting others, after the side walls and the columns of the latter have been removed.'

1892, App. 24 October, Acc. 15 July 1893: GB Patent No.19,070
Carl Lührig:
Improvements in Locomotive Engines or Vehicles for Street Tramways. Summarised as '1. A locomotive vehicle for street tramways wherein the motor engines, the driving gear and all accessory parts are enclosed within a vehicle having the same outward appearance as an ordinary tramway car. 2. In a locomotive vehicle such as is referred to in the first claim, effecting the change of speed and direction of motion by means of a spur or bevil wheels provided with clutches, to which wheels the motion of the motor shaft is imparted through other toothed wheels or worm wheels, substantially as described. 3. In a locomotive vehicle such as is referred to in the first claim, the arrangement of a curved buffer plate, extending over the whole width of the vehicle and provided with an elastic covering, the said plate being made to bear with rods against springs. 4. In a locomotive vehicle such as is referred to in the first claim, the use of a pivoted coupling rod which can slide in slots in the framing so as to assume varying angular positions.'

1893, App. 22 February, Acc. 23 December: GB Patent No.3942
Carl Lührig:
Improved Friction Clutch Apparatus. Summarised as 'A frictional clutch apparatus wherein two loose discs are brought into close frictional contact with an intermediate disc fixed on the shaft, by means of screw spindles that are rotated by the action of a sliding sleeve to which they are connected by rods.' One of a number of clutch designs proposed by Lührig.

1893, App. 12 June, Acc. 14 April 1894: GB Patent No.11,506
Carl Lührig:
Improvements in Locomotive Tramway Vehicles. Summarised as 'In a locomotive tramway vehicle of the kind described in the Specon. No.15841 of 1892, arranging reservoirs for

gas and water below seats and under framing...'. Despite being granted his patent, neither of these proposals described by Lührig was actually original, as can be seen from several other patents on this list.

1894, App. 29 May, Acc. 24 May 1895: GB Patent No.10,452
Eustace Fitzmaurice Piers, Bart:
Improvements in or connected with Locomotive Engines or other Engines subject to Intermittent Work or Varying Loads, more especially intended for Use with Tramway Engines. The first of a series of designs patented by Piers between 1894 and 1896 in which he proposed to use a small gas engine is used to pressurise four pumps linked to an hydraulic motor to drive the tramcar. He also suggested using an engine fuelled by 'other motive agents' – or even steam. Summarised as 'In connection with locomotive engines or other engines subject to intermittent work or varying loads with combination with the primary motor and two or more pumps operated by the said primary motor for pumping water or other fluid which acts as the motive agent for operating a secondary motor, for propelling a tramcar or for doing other work, means by which any one of more of the said pumps may be caused to cease to act as a pump whereby increased power of the secondary motor may be obtained at the expense of speed substantially as hereinbefore specified.'. The others were No.10,451, 29 May 1894; No.15,560, 15 August 1894 and No.30,065, 31 December 1897.

1895, App. 17 September, Acc. 12 September 1896: GB Patent No.17,282
Frederic O'Connor Prince:
Improvements in Means of Propelling Vehicles by Internal Combustion Motors. Prince's motor was designed to run on a mixture of air and gas, or vaporised oil and to have variable gears by the use of belts and gear cones. A unique feature of his patent was in the means of turning the engine at either end of a tramway to ensure the driving position faced in the direction of travel. Summarised as 'I fit the car with a supplementary wheel at or near each corner, and at an angle taken from the centre of the car, and so arranged that each wheel can be screwed or forced down upon the ground or a circular track, to bodily lift the car to raise the usual bearing wheels of

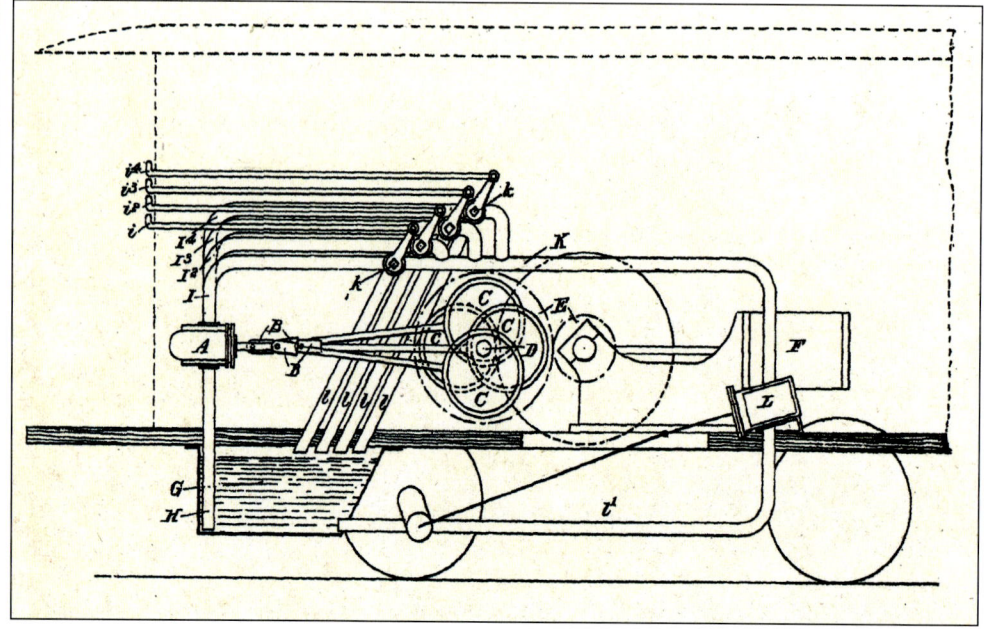

The elaborate layout of Eustace Fitzmaurice Piers' hybrid tramcar engine, showing the four hydraulic pumps – powered by a small gas engine – control of which allowed the driver to adjust the traction possible with the engine. The pumps effectively worked as gears.

their rails. To facilitate the turning of the car at the terminus of each track I arrange a circular rail crossing the rails for the supplementary wheels to bear upon when screwed down and secured, so that the car is supported by the supplementary wheels on this circular track, when the car can be easily turned round by the driver.' None of the drawings accompanying his patent illustrate this feature.

1896, App. 11 June, Acc. 11 June 1897: GB Patent No.12,901
Howard Lane:
Improvements in Machinery or Apparatus for the Propulsion of Tram-cars and other Vehicles. Summarised as '1. In machinery or apparatus for the propulsion of tramcars or other vehicles, the combination—upon the vehicle—of a gas engine or other motive power engine, pumps, accumulator, hydraulic motor with means for varying the stroke of its plungers or converting it from a motor to a pump, all arranged and operating substantially in the manner described and illustrated. 2. In combination with the motive power engine, pumps, accumulator, and hydraulic motor, the crankpin driven by the plungers of the motor, and adjustable…'

1903, App. 11 September, Acc. 11 December 1903: GB Patent No.27,370
Lucas-Girardville, Paul Nicholas, and Mékarski, Louis:
Combined Internal Combustion and Compressed Air Motors. Having patented the compressed air tramcar almost thirty years earlier, Mékraski teamed up with Lucas-Girardville to propose a tramcar where an internal combustion engine – which he described as an 'explosion motor' was used to drive a compressor to deliver air to the 'air motor. They also proposed using the waste heat from the internal combustion engine to generate steam which was mixed with and warmed the compressed air as it entered the cylinder. The fuel specified in the patent was to be 'Water Gas', a combustible mixture of carbon monoxide and hydrogen produced by passing steam under pressure over or through red-hot coal. The vehicle did not enjoy the success of Mékarski's compressed air tramcars.

Despite Frederic Prince's patent making much of the procedure by which the tramcar could be rotated at the terminus, none of the illustrations demonstrated that capability – concentrating instead on the geared chain drive.

A CHOICE OF GASES

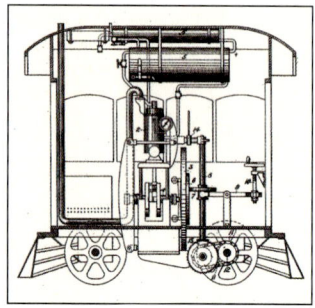

One of several versions of the Connelly Motor which used an engine fuelled with vaporised naphtha to haul former horse cars.

For the many pioneering inventors of gas engines and the transport systems they believed would use them, there were a number of potential fuels to choose from – some markedly more efficient than others. In the closing years of the nineteenth century, however, the ready availability of 'town gas' – coal gas – in many towns and cities made the choice a relatively straightforward one for most engineers and inventors.

Today, there is a well-established national gas grid already distributing 'natural gas' – predominantly methane – throughout the country. 'Green' biomethane is already being fed into the network from a small but growing number of treatment plants, and if that was replicated across the country, refuelling gas vehicles with biogas has the potential to become easier and easier – if only government policy accepted that the internal combustion engine has a 'green' future.

When William Murdoch, or Murdock as he later preferred to be known, first experimented with his flammable gases – the major use for which he predicted would be illumination – he was following on from centuries of experimentation by others. In his paper read to the Royal Society on 25 February 1808, he established the timeline of his investigations.

> 'It is now sixteen years since, in a course of experiments I was making at Redruth in Cornwall, upon the quantities and qualities of the gases produced by distillation from different mineral and vegetable substances. I was induced by some observations I had previously made upon the burning of coal, to try the combustible property of the gases produced from it, as well as from peat, wood, and other inflammable substances. And being struck with the great quantities of gas which they afforded, as well as the brilliancy of the light, and the facility of its production, I instituted several experiments with a view to ascertaining the cost at which it might be obtained, compared with that of equal quantities of light yield by oils and tallow.'

But by the time that paper was read on his behalf – Murdoch was not even invited to attend – others had recognised that the thermal properties of gas would turn out to be of much more enduring importance than its qualities as an illuminant.

Murdoch's paper did offer some detail on the production methods he was using, and the set-up of heating coal in a closed retort – initially a large kettle-like container – had already been established, as had bubbling the gas through water to 'clean' it.

'Coal gas', 'town gas', or just 'gas' – the names by which it was known in Victorian and Edwardian Britain – is actually a mixture of a number of gases, some flammable, others non-flammable. The mixture usually consists of hydrogen (H_2), methane (CH_4), carbon monoxide (CO), carbon dioxide (CO_2), Nitrogen (N), and water vapour (H_2O). Over the years, scientists and inventors would come to recognise that varying the amount of oxygen present as the coal was being heated played a significant role in establishing the character of the gas produced.

Murdoch first successfully demonstrated the flammable qualities of coal gas around 1790, announcing his 'discovery' in 1792, but as early as the 1780s, others were experimenting with the production of gases derived from other sources, and what few understood in those early days was that the gases they were producing were quite different, both chemically and in their thermal characteristics.

At the time gas engines were being developed, there were a number of different gases available which could fuel them.

The earliest was probably 'water gas' – often attributed to the experiments of the Italian chemist Felice Fontana about 1780, but which really only came to prominence in the late 1820s. Then came 'town gas', followed by 'producer gas'. All three are cited in the gas engine patents collected for this project. Chemically, all three were quite similar in many respects, but the manner of their manufacture – and their potential effectiveness for use in gas engines – differed considerably.

Water gas consisted chiefly of carbon monoxide and hydrogen with small amounts of methane, carbon dioxide and nitrogen, and was made by blowing air and steam sequentially through white-hot coke, creating a gas with a higher nitrogen content than coal gas, and thus a lower calorific value.

In the manufacture of producer gas, air was blown through red hot coke or anthracite, resulting a gas with an even higher nitrogen content – as high as two-thirds by volume of the gas produced, giving it an even lower calorific value. But it had the benefit of being as ideally suited to production on a small scale as a larger scale. Thus, producer gas engines often came with their own transportable 'gasworks' rather than being reliant upon a fixed piped supply.

In the manufacture of coal gas, uniquely the coal was not burned, but heated in a retort in a highly oxygen-depleted environment. The resulting gas had a much higher calorific value, and the 'residue' from the process was coke, which turned out to be at least as valuable as the gas itself, playing a key role in the growing demand for iron and steel.

The relative calorific values of coal gas, water gas and producer gas would be measured as approximately 550BTU, 300BTU and 137BTU per cubic foot respectively – or 18,540, 10,110 and 4,617 kilojoules per cubic metre in metric terms.

As far as gas tramways were concerned, access to refuelling stations was key, and with town gas that meant laying pipework from the local gasworks to appropriate 'charging points', either at the route's termini, or additionally at intermediate points with longer lines. But as the tramways which used gas traction were all in towns which already had a piped supply *in situ*, the only additional cost was in the construction of compression stations to refill the tramcars' tanks.

The other gas mentioned in late nineteenth century patents was naphtha vapour, made by warming liquid naphtha – which has a very low boiling point, typically under 30°C – and which can be made from a variety of hydrocarbon sources. Back then it was primarily recovered from either distilled coal tar or produced as a by-product of the refining of petrol (gasoline). Naphtha is a highly volatile and highly toxic chemical compound. Burned in the Connelly Motor, it typically generated twice the energy as town gas, and almost as much as today's natural gas – methane – and biomethane.

While all four gases were cited in patents as having potential as fuels, only coal gas and naphtha fuelled engines were ever used on commercial tramways.

BIBLIOGRAPHY

The early 1880s saw a massive expansion of Britain's tramways, with each new route – like the railways earlier in the century – requiring enabling legislation. Under the general heading 'Legislation, Projects of', the 1882 edition of Charles Dickens Jr's *Dickens's Dictionary of London* devoted several pages to detailing the new tramway routes being proposed and awaiting legislation. By the 1888 edition, that had been replaced by three pages of tramway routes and fares.

Abel, P. H. and McLoughlin, I., *Blackpool Trams: The First Half Century 1885–1932*, Oakwood Press, 1997

Anon. *De Gasmootortram volgens het Systeem Lührig*, The Gas Traction Company Departm. Netherlands, 1896

Barbet, Louis-Alexandre, *L'air Comprimé Appliqué À La Traction Des Tramways*, Baudry et Cie., 1896

Brockhaus, Friedrich Arnold: *Brockhaus' Konversations-Lexikon*, 14th Edition, 1894

Bucknall Smith, J., *Treatise upon Cable or Rope Traction*, Offices of Engineering, 1887

Clark, D. Kinnear, *Tramways – Their Construction and Working with special reference to the Tramways of the United Kingdom*, Crosby, Lockwood & Son, 1894

Clark, D. Kinnear, *The Steam Engine – a Treatise on Steam Engines and Boilers*, Blackie & Son, 1890

Dawson, Philip, *Electric Railways and Tramways, their Construction and Operation. A Practical Handbook*, 'Offices of Engineering' 1897

Dickens, Charles, Jr., *Dickens's Dictionary of London*, Macmillan & Company, 1882 and 1888 editions

BIBLIOGRAPHY

Donkin, Bryan, *A Text-Book on Gas, Oil and Air Engines, or Internal Combustion Motors without Boiler*, Griffin & Company, 1894

Gladwin, David, *A History of the British Steam Tram, Volume 7*, Adam Gordon, 2010

Gladwin, David, *Horse and Steam Trams of Britain*, Nostalgia Road Publications, 2014

Gray, Edward, *Trafford Park Tramways*, Oakwood Press, 1964

Green, Oliver, *Rails in the Road: A History of Tramways in Britain and Ireland*, Pen & Sword Transport, 2016

Griffiths, John, *The Third Man, the life and times of William Murdoch, inventor of gaslight*, Andre Deutsch, 1992

Hannavy, John, *Transporter Bridges – an illustrated history*, Pen & Sword Transport, 2020

Höse, Dietmar, *Fahre mit Gas*, D. Höse 2016

Hutton, Frederick Remsen, *The Gas-Engine a Treatise on the Internal-Combustion Engine Using Gas: Gasoline, Kerosene, Alcohol, or Other Hydrocarbon as Source of Energy*, Chapman & Hall, 1909

Jacobi, Sébastien, *Neuchâtel en Tram 1890–1990*, S. Jacobi, 1989

Jenniskens, Antoon Hendrik, *Pak de bus*: Stichting Historische Reeks, 1995

King, Robert, *Neath Through Time*, The History Press, 2010

Kreschnak, Werner, *Geschichte der Dresdner Strassenbahn: Geschichte des VEB Verkehrsbetriebe der Stadt Dresden (1872 bis 1975)*, Tribüne, 1981

Lamming, Clive, *Paris Tram*, Parigramme, 2003

Langen, Arnold, *Nicolaus August Otto: Der Schöpfer des Verbrennungsmotors*, Frankh'sche Verlaghandlung, 1949

Lucke, Charles Edward, *Gas Engine Design*, D. Van Nostrand Company, 1905

Mayhew, Henry, *London Labour and the London Poor*, Griffin, Bohn & Co, 1861–2

Mehrtens, August Christian, *Gas Engine Theory and Design*, John Wiley & Sons, 1909

Richard, Gustave, *Les Moteurs à Gaz*, Veuve Ch. Dunod, 1885

Roberts, David, *Neath and Port Talbot Remembered*, Breedon Books, 2001

Sittauer, Hans L., *Gebändigte Explosionen: Nicolaus August Otto und sein Motor*, Transpress Verlag für Verkehrswesen, 1972

Smiles, Samuel, *Lives of the Engineers Volume 3*, John Murray, 1862

Southey, Robert, *Journal of a Tour in Scotland in 1819*, John Murray, 1929

Stretch, E. K., *The Tramways of Wigan*, Manchester Transport Museum Society, 1978

Tangye, Richard, *"One and All", an Autobiography of Richard Tangye*, S. W. Partridge & Company, 1902

Tookey, William Alfred, *The Gas Engine Manual: A Practical Handbook of Gas Engine Construction and Management*, P. Marshall & Co., 1908

Turner, Brian, *Lytham Trams: The Blackpool, St Annes & Lytham Tramway*, B. Turner, 2020

Turner, Keith, *The Directory of British Tramways*, Patrick Stephens Ltd., 1996

Voice, David, *The Definitive Guide to Trams (including Funiculars) in the British Isles*, Adam Gordon, 2005

Voice, David, *Explosive Power on Tramways in the British Isles: The story of tramways using internal combustion engines*, Adam Gordon, 2017

Whitcombe, H. A., *The History of the Steam Tram*, Adam Gordon, 2000

Winchester, Simon, *Exactly – how precision engineers created the modern world*, William Collins, 2018

Crossovers and points were used to give controlled access to sheds and depots. Although tramways in Britain ran on a number of gauges, the profile of their rails has been largely unchanged from Loubat's 1853 design.

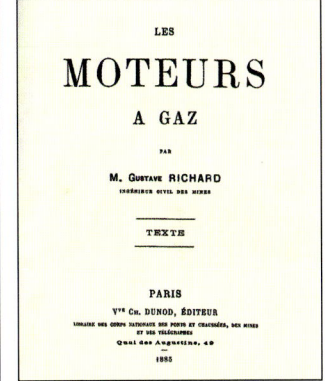

The title page of *Les Moteur à Gaz*, Gustave Richard's contemporary account of the development of the gas engine and its application to tramcars. The book is an informative document, its text serving as an accompaniment to the separately published portfolio of plans and diagrams from inventors and patent-holders all over the world.

ACKNOWLEDGEMENTS

The rebuilt Neath Tramways gas tram at Cefn Coed with both its engine access doors open. The flywheel was essential technology in the evolution of the gas engine. In the original design, the whole engine assembly could be slid out on to a trolley and replaced, allowing maintenance to be carried out without disrupting services. Today, flywheel technology is still being used in many innovative ways.

This project has depended upon the generosity of many people in following up leads on my behalf when access to collections during the pandemic permitted it. Special thanks go to Steven Campion at the British Library, Geoff Challinor OBE at the Anson Engine Museum, and Brent Efford at the New Zealand Tramway Historical Society, whose searches in the NZ Archives in Wellington unearthed the missing illustrations for Patent 681. Thanks also to: David Voice; Keith Davies and Owain Williams at Cefn Coed Museum; Keith Tucker at the Neath Antiquarian Society; David Edwards at the Atwell-Wilson Motor Museum; Marietje Ruijgrok at the University of Delft; Phil Hulme-Jones; Robert Haley; Ashley Birch; Edward Dawes; Martin Dibbs; Terry McElarney; Paul Abell; John Prentice; Fred Starr; Peter Wade; Colin Withey; Peter Waller; Geoff Burns at Birmingham Archives; Adele Heagney at the St. Louis Public Library; James Dalton/Australian Railway Historical Society Archives; Gael Newton, Mal Rowe, Marcel Safier, and Brian Weedon also in Australia; Christine Hall at the Waitaki Museum and Archive, New Zealand; Alice Meads at the New Zealand Archives; Dave Hinman; Sébastien Jacobi (author of *Neuchâtel en Tram*) and Laurent Maeder at the Musée du Tram, Neuchâtel, Switzerland; Neeltje Wessels, A.H. Jenniskens and John Kerkhofs in the Netherlands; Brad Read at TIG/m; Alexander James and Peter Stubbs in Edinburgh; Beverley Nielson, Ultra Light Rail Partners Ltd; Paddy Fawcett, Invizio Product Design; Christopher Maltin of Biomethane Ltd, and Technical Director of ULRP; James Skinner; John Parry MBE at Parry's People Movers; Summerlee Museum of Scottish Industrial Life, and Ulrike Krautz at the Verkehrsmuseum in Dresden.

The information they all passed on and the facilities they extended to me have helped develop a much more complete history of these unusual vehicles. Any mistakes, however, are all mine. Thanks also to my cousin Chris Masterton for his photography in Edinburgh and, as ever, to Kath my wife for her continuing support and encouragement.

The illustrations are from the author's collection and two private collections except: Chris Masterton p49; T. P. Lugton Archive/Alexander James p36 bottom; Geoff Challinor OBE p60; Crossley Archive/Anson Engine Museum p167; Robert Haley p129–131, 134–5, 137, 138 top; *Scientific American* p80 top; Private Collection, Switzerland p85–7; The National Archives of Australia, Document NAA: A13128, 554, p95–6; Archives New Zealand, Patent Specification 681 ref:ABPJ7396W3835 Box 14, Engines (Air, Gas & Oil) 1 ref:ABPJ7746W5650988 p102, 105–6; Australian Railway History Society Archives p34 bottom; Historic Centre of Limburg Fotocollection GAM (image Nos.1275, 11374, & 31885) p90 bottom, 91 top, 92; David Voice p128, 145; Neath Antiquarian Society p147, 150, 153, 154 top, 156–7, 158; Keith Davies p121 top; The Commercial Motor Archive p161. David Bayes Collection p138 bottom, p140 top; © TfL from the London Transport Museum collection p117 top; Neath Antiquarian Society and Neath Port Talbot CBC p157; Ultra Light Rail Partners/Invizio p172, 175; Wessex Water/GENeco/Julian James 169 top; Paul Abell/ULRP p174; Brad Read TIG/m p177–9; The letter on p111 is reproduced courtesy of the Anson Engine Museum

INDEX

Abel and Imray, Patent Agents, 71, 74
Accrington and Haslingden Tramways, 28
Accrington Corporation Steam Tramways, 169
Aldred, Joseph Hulme, 29
Alford & Sutton Tramway, 34
Alton Court Waterworks, 54
American Air Power Company, New York, 139
Amsterdam Tramways, 84
Annibas, 8
Aruba, Lesser Antilles 9, 177-8
Ashbury Railway Carriage and Iron Company Ltd, 7, 9, 119-121, 123-4, 126-7, 146-8, 155
Austin Seven 'Chummy', 158
Australian Railway History Society, 34

Baker, J. Alan, 139
Barber, John, 180
Barnes, Benjamin and Danks, John 8, 94-8, 101
Barnett, John, 180
Barnsley, Harry, 156-7, 160
Beamish, the Living Museum of the North, 17, 170
Beckton Gasworks, 55, 65
Beattie, William Hamilton, 49
Beau de Rochas, Alphonse-Eugene, 182
Beddoes, Frank, 158-160
Belfast Newsletter, The, 31
Belfast Street Tramways Company, 24
Benz engines, 86
Bertrand, M., Compagnie Parisienne, 85
Beyer-Peacock, 32
Biomethane, 9, 173-9, 197
Biomethane Ltd., 174
Birmingham Cable Tramway, 43

Birmingham Central Tramways Company, 29
Birmingham Steam Trams, Remembrance postcards, 35, 164
Blackburn Corporation Tramways Company Limited, 28
Black Country Living Museum, 65
Blackpool, St. Anne's and Lytham Tramway Company, 81, 109, 123, 127, 129
Blackpool, St. Anne's and Lytham Tramways Company Limited, 128-32, 134 138-9, 142, 145, 153, 171
Blackpool tramways, 92, 168
Blackstone Edge, 13
Blessing, Oskar, 8, 70, 72, 191
Boston Gasworks, 53
Boult, Alfred Julius, Patent Agent, 74
Boulton and Watt, 56-7
Bradshaw, Edwin & Son, auctioneers, 146
British Gas Traction Company, The, 89, 119, 123, 125, 127-8, 130-133, 137, 139, 141-4, 151, 155, 174
British Motor Museum, 15
Brockhaus' Konversations-Lexicon 39
Brown, David, 161
Buck, M. M., Manufacturing Company, 97

Caledonian Canal, 11
Campbell Gas Engine Company, 63
Cartmell, W. H., tram driver, 141
Coutchouc Gas Bags, 120
Capitaine, Emil 8, 191
Carrick, William C., 107-8
Cassier's Magazine, 8, 21, 69, 80, 120
Cefn Coed Colliery Museum, 7, 148, 150
Charlier, Albert, 81
Clifton Hill to Alphington Railway, 94, 96-7

Coal gas, 196
Cochrane, Archibald, Earl of Dundonald 53
Coleridge, Samuel Taylor, 11
Commercial Motor, The, 161, 163
Compagnie Parisienne, 85
Connelly Gas Motor Company, The, 8, 82, 86, 117-18, 196-7
Connelly, John S., 99, 115-18, 188-9, 191
Crich Tramway Village, 50
Crossley Brothers, 61, 63, 86, 113, 115, 125, 137, 167
Crossley, Francis William, 66, 86, 110-13, 115, 165
Cross Town Electric Railway Company, St. Louis, 99
Croydon and Thornton Heath Tramways Company, 9, 80, 117-19, 135
Curtis, William Joseph, 20, 181

Daimler, Gottleib, 8, 61, 185-6
Danks, John & Company, 94
Danks, John and Barnes, Benjamin, 8, 94-8, 101, 187
De Gasmotortram volgens het Systeem Lührig, 8
Deprez, Marcel, 67
Derby Horse Cars, 164
Dessau Tramway, 8, 78, 80, 82, 85, 93, 119, 134
Deutsche Gasbahn Gesellschaft, 77, 85
Dickens, Charles Junior, 198
Dickens's Dictionary of London, 23
Dick, Kerr & Company, 29, 34, 46, 65, 162, 168
Didcot Railway Centre, 30
Doha, Qatar, 177-9
Dresden tramways, 82, 92-3, 134
Dubai tramcars, 178
Dunlop, John Boyd, 15
Dunlop, John Macmillan, 111
Dyson, 8

Eades Patent Reversible Horse Car, 33
Edinburgh and District Tramways Company Limited, 36-7, 49
Edinburgh Northern Tramways Company Limited, 49
Edison Electric Light Station, Holborn, 165
Electrical Power Storage Company, 51
Electric Railway & Tramway Carriage Company Limited, the, 46
Eerste Nederlandsche Gastractie Maatschappij, 89

Falcon Engine and Car Works Limited, Loughborough, 19, 29-30, 34
Fakenham Gasworks, 52-3, 55-7
First Gas Traction Company of the Netherlands, 89
Fontana, Felice, 197
Fox, Walker & Company, 33

Gas, Light and Coke Company, 57
Gasmotoren-Fabrik Deutz, 61, 64, 77, 124, 145, 155
Gas Traction Company, The, 8, 78, 80-84, 120, 124-5, 133
Gas Traction Company Departm. Netherlands, 78-9, 107
Gatwood, Walter, 124
Gaunt, William Henry, 146
GENeco, 173
Giant's Causeway, Portrush & Bush Valley Railway, 28
Gillespie, Thomas, 133
Gilliéron and Amrein, 85-7, 93
Glasgow Corporation Tramways, 170
Gnom Engines, 86
Grantham, John, 27-8, 33
Great Eastern Railway, 30
Great Orme Tramway, 46, 48, 50
Green, Thomas & Son, 28-30, 32
Griffin, Samuel & Company, 65
Gross-Lichterfelde Tramway, 83

Harper, Walter Andrew, 101-6, 186
Harrison, Frederick James 8, 186-7

Hereford Waterworks, 54, 63
Hersey, Thomas, 83, 136
Highgate Cable Tramway, 36-7, 47
Hillman 'Imp', 158, 160
Hirschenberg Valley, 88-9
Holt, Henry Percy 8, 66, 86, 108, 110-13, 115, 124-5, 133, 135, 137, 155, 165, 182
Holyoake, Percy, 127, 133-7, 144-6
Home House Foundry, Pemberton, 27
Hooley, Ernest Terah, 144
Hot Tube Gas Engine, 63, 65
Hughes, Henry & Company, 19, 24-5, 30, 34
Hughes, John, 37, 39-40, 42
Hughes & Lancaster, Wrexham, 37, 39-40
Hutchinson, Major Charles Scrope, 80, 114-5, 118-19
Hydrogen fuel cells, 176-8

Illustrated Newspapers Limited, 51
Institution of Civil Engineers, The, 137
Ipswich Tramways Company, 134, 136-7, 168

Jelenia Gora, 87-90

Karachi Tramways, 161
Kelham Island Museum, Sheffield, 63
Kemper, A., 89, 93
Kennedy, Professor Alexander, 119
Kinnear Clark, D., 89
Kitson & Company, Leeds, 27, 29
Korting Brothers, 8
Krauss, Conrad, Munich, 66-7, 96, 113, 184
Krauss, Augustus and Son, Bristol, 30, 147, 149, 151-2

Lancaster Carrige Works, 7
Lancaster Railway Carriage and Wagon Company Limited, 7, 123, 145, 149
Langen, Eugen, 59, 81
Lane, Howard, 43-, 195
Lavezzari, M. A., 80, 82-3, 92, 104, 107, 124

Legros, Lucien Alphonse, 139, 146
Lenoir, Jean Joseph Etienne, 58-9, 182
Levy, M., Compagnie Parisienne, 85
Leyland Motors, 161
Liquefied Natural Gas, LNG, 175
Lithium batteries, 176
Lobenhofr, 8
Locomotivfabrik Krauss, Munich, 66-7, 96, 113, 184
London and Westminster Light and Coke Company, 57
London County Council, 44-5
London General Omnibus Company, 14
London Labour and the London Poor, 13
London Street Tramways Company, 23
London Tramways Company, 23
Loubat, Alphonse, 17, 19, 22, 181, 199
Lucas-Girardville, Paul Nicholas, 195
Lührig, Carl, 8, 67-80, 82-3, 85-7, 93, 103, 108, 113, 120, 146, 192-4
Lührig-Holt System, 125, 152

Maastricht Tramway, 88-92
Macadam, John Loudon, 12-13
MacBrayne, David, 11
McGee, John, 189-90
McGhee, Richard, 189-90
McNay, Thomas Fothergill 8, 186-7
McNeill, Andrew, 187-8
Manchester Carriage & Tramways Company, 33
Manchester Ship Canal Company, 145, 147
March, William, 8, 191-2
Marchant and Wrigley, 67
Marindin, Colonel Sir Francis, Board of Trade, 149
Mason, Freank H., 21
Matlock Cable Tramway, 47, 50
Matthews, James, 33
Maxim, Hiram, 44, 47
Maybach, Wilhelm, 61
Mayhew, Henry, 13-14
Mékarski, Louis, 38-42, 176, 182, 195
Merryweather & Sons, 27, 29, 34
Montclar, Jean Marie Armand, 8, 107-8, 186

Morani, 8
Morecambe Tramways Company Limited, 161-3
Motorenfabrik Oberursel, 86
Moulton, J. Fletcher, Q.C., 137
Muirtown Locks, 11
Murdoch (or Murdock), William, 53, 56-7, 196

Naphtha, 117, 197
National Gas Engine Company, 54
Neath Borough Council Training Agency, 156-7
Neath Corporation Traways, 92, 121-23, 145, 147-61, 200
Neath Gas Traction Company, 152Neilson and Company, Glasgow, 55
Neuchâtel and Saint-Blaise Tramway, 86-7, 93
Nichols, George Benjamin, 8, 191-2
Noble, Jay, 97-100, 190
Northampton Horse Cars, 164
North Metropolitan Tramway Company, 23
North London Tramways Company, 34
North Staffordshire Tramway, 34
Nottingham and Districts Tramways Company Limited, 24-5

Oamaru-Moeraki Railway, 101
Oamaru Tramway Act, 1876, 101
O'Kelly, James Morris, 107, 109, 192
Oldbury Carriage and Wagon Works, 27
Otto-cycle gas engine, 59-62. 64, 76, 81, 92, 94, 107, 123, 154-5, 166-7, 169
Otto-Langen Engine, 112
Otto, Nicolaus, 59-62. 64, 76, 81, 92, 112, 124, 183, 190
Otto Gas Engine Works, Philadelphia, 62
Oystermouth Railway, 18

Parry People Movers Ltd., 174
Patents
    Barber, John, 180
    Barnes, Benjamin, and Danks, John, 8, 94-8, 187
    Barnett, John, 180
    Beau de Rochas, Alphonse-Eugene, 182
    Blessing, Oskar, 8, 70, 72, 191
    Carrick, William C., 107-8
    Connelly, John S, 189, 191
    Connelly, John S, and Connelly, Thomas E., 188
    Crossley, Francis, and Crossley William, 183-4
    Curtis, William Joseph, 20, 181
    Daimler, Gottlieb, 185
    Danks, John and Barnes, Benjamin, 8, 94-8, 187
    Harper, Andrew, and Rock, John William, 101-6, 186
    Hilton, Matthew & Johnson, Samuel, 184
    Holt, Henry Percy, 182
    Holt, Henry Percy, and Crossley, Francis William, 110, 112, 184-5
    Hughes, John, 37, 39-40
    Krauss, Conrad, 66, 113, 184
    Lane, Howard, 43, 195
    Lenoir, Jean Joseph Etienne, 58, 182
    Loubat, Alphonse, 181
    Lucas-Girardville, Paul Nicholas and Mékarski, Louis, 195
    Lührig, Carl, 8, 67-80, 82-3, 85-7, 93, 192-4
    McGhee, Richard and McGee, John, 189-90
    McNay, Thomas and Harrison, Frederick 8, 186-7
    McNeill, Andrew, 187-8
    Mékarski, Louis, 38, 182
    Montclar, Jean Marie Armand, 107-8, 186
    Nichols, George Benjamin and March, William, 8, 191-2
    Noble, Jay, 97-100, 190
    O'Kelly, James Morris, 107, 109, 192
    Otto, Nicolaus, 183-4, 190
    Prince, Frederic O'Connor, 194-5
    Purssell, John Roger, 114, 185
    Quick, Joseph and Joseph, 66-7, 96, 113, 186
    Rock, John William, and Harper, Andrew, 101-6, 186
    Stevens, Emmanuel, 189-90
    Wilkinson, William, 31-2
Penrhyn Castle, 55
Philips and Lee, Salford, 57
Pierper, Carl, 66
Piers, Eustace Fitzmaurice, 8, 194
Pilling, Alderman Abraham, 124-5
Preston Corporation Tramways, 168
Prince, Frederic O'Connor, 194-5
Producer Gas, 54, 197
Provincial Gas Traction Company, 152-3, 155
Purrey Steam Tram, 30
Purssell, John Roger, 114, 185

Quick, Joseph and Joseph, 66-7, 96, 113, 186

Rechenzaun, Anthony, 51
Renewable Natural Gas, RNG, 175
Retort Houses, 52-3, 55, 58
Rhyl Tramways, proposed, 121
Ribble Transporter Bridge, proposed, 124
Richard, Gustave, 66-7, 110-11, 199
Rochdale Corporation Tramways, 171
Rock, John William, 101-6, 186
Rowlandson, Thomas, 56-7

Saint Petersburg Electric Tramway, 155
Sandy & Potton Light Railway, 30-31
Scania, 173
Schleicher, Schumm & Company, Philadelphia, 62
Scientific American, The, 80, 85
Serpollet, Léon, 30, 83
Sheffield Tramways, 168
Siemens, Charles 83, 165-7, 169
Société des Ingenieurs Civil de France, 92
Soho Foundry, 57
Southey, Robert, 11
Southport and Lytham Tramroad Act, The, 124
Southport and Lytham Tramroad Company, The, 125
Stephenson, Robert & Company, 111

Stevens, Emmanuel, 8, 189-90
Stoll, Carl, 77
Stratford Locomotive Works, 30
Sunderland Corporation Tramways, 170
Systeme Mékarski, 38-42

Tangye Brothers, 43, 64-5, 146
Telford, Thomas, 11-12
The Age, Melbourne, 94
Thomas, William Luson, 51
Thomson, Robert William, 15
TIG/m Modern Street Railway, 177-8
Tower Bridge, London, 13
Town Gas, 196
Traction Syndicate Limited, The, 120
Trafford Park Tramway, 139, 141-6
Tramway a Gaz, 80-81
Tramway Funiculaire de Bellevile, The, 50
Tramways Act, 1870, 20, 165
Tramways Museum, Crich, 32

Ultra Light Rail Partners Ltd., 173-4
United Electric Car Company, 161-2, 171

Van der Zypen und Charlier, Cologne, 77, 81
Van der Zypen, Ferdinand, 81
Vapourised Naphtha, 117, 197
Victorian Railways, Australia, 94
Volk, Magnus, Electric Railway 165-7

Wallace, Roger W., Q.C., 137
Wantage Tramway, 27, 30, 33
Water Gas, 197
Webster's Dictionary, 17
Wessex Water, 173
Westland Steam Tramway Company, 89
Westlandsche Stoomtramweg Maatschappij, 89
West Manchester Light Railway Company, 139, 147
West Metropolitan Tramways Company, 51
Weymann & Company, Guildford, 117
Widnes-Runcorn Transporter Bridge, 61
Wilkinson, William and Company, 27, 29, 31-2
Wisbech & Upwell Steam Tramway, 30-31
Wolverhampton Corporation Tramways, 171
Wolverton & Stony Stratford Tramway, 30
Wordsell, Thomas, 30
Wordsworth, Dorothy, 11
Wordsworth, William, 11
Worsley Mesnes Company, Wigan, 7

York Tramways Company, 24-5
Youth Training Scheme, 159